# HEAT

ALSO BY PETER A. MICHEELS

BRAVING THE FLAMES

THE FIRE INVESTIGATORS

AND

THEIR WAR ON ARSON

# H E A T

AND

MURDER

·

PETER A. MICHEELS

ST. MARTIN'S PRESS NEW YORK

EDITOR: JARED KIELING
PRODUCTION EDITOR: RICHARD KLIN
COPYEDITED BY ASHER JACOBS
DESIGN BY JUDITH A. STAGNITTO

Library of Congress Cataloging-in-Publication Data

Micheels, Peter A.
    Heat : the fire investigators and their war on arson and murder /
    Peter A. Micheels.
        p.    cm.
    ISBN 0–312–04848–3
    1. Fire investigation—New York (N.Y.) I. Title.
TH9180.M53 1991
362.2'5—dc20                                          90–27342
                                                         CIP

First edition: July 1991
10  9  8  7  6  5  4  3  2  1

This book is dedicated to the more than 752 FDNY firemen

who have lost their lives in the line of duty,

and to those fire marshals

who risk their lives apprehending arsonists and other criminals.

It is also dedicated to my father,

whose unrealized dream was

to be a detective.

# ACKNOWLEDGMENTS

This book was completed with the help of some extraordinary people. I first want to thank those men who, with complete honesty, shared their stories with me and allowed me to partake in the camaraderie, humor and adventure of the Bureau of Fire Investigation. I would also like to thank the following people from the BFI who gave me an understanding of what it is to be a fire marshal in New York City: Charles Wagner, Arthur Crawford, Tom Mulvey, Bernie Casey, Mike DiMarco, Terry Cullen, John Gallagher, Ed Peknic, Ronnie Ferlazzo, Donald Forster, Emil Harnischfeger, Ed Wysocki, Chris Heesch, Cecil P. Maloney, Jr., Luis Garcia, Ed Fanuzzi, George Tufte, Donald Washington, John Masterson, Jim Kelty, and Gene Fraher. The following people get my thanks for providing me with important background information: Fire Fighter Kenny Kearney, E. 60; Captain George Eysser, L. 6; Dispatchers Herb Eysser and Bruce Brenner, Manhattan Communications Office; Assistant Commissioner John Mulligan; Fire Commissioners Joseph Bruno and Carlos Rivera; and Ms. Gloria Sturzenacker, Managing Editor, WNYF.

•

My appreciation also goes to Anthony Guerriera, Frank Cantalupo and Tom McNulty, Forensic Photo Unit; Lt. James Curran, President, N.Y. Fire Fighters Burn Center Foundation; Chief Bart Mitchell and FDNY chaplain Fr. Julian Deeken, who first got me involved with the marshals.

Mrs. Cecilia Anzalone's transcription was a critical step in getting this book into print. Andrew Anzalone provided a vital link in the process of turning audio cassettes into typed transcripts.

I would like to mention a special word of thanks to Hal Freitag for the diagram of the tenement, and to Steven Spak, Bob Athanas, Maggie O'Bryan, and Harvey Eisner for their photographs.

The Bellevue Hospital Crew: Belinda Greenfield, M.Ed.; Arthur Anderson, M.A.; Gloria Miele, M.A.; Louis Cuoco, A.C.S.W.; Kathleen Ready, M.S.W.; Bernard Salzman, M.D.; Fred Covan, Ph.D.; and Morgan Hurley, J.D., get my deep gratitude for all of their help and support.

I also want to thank my agents Beth Waxse and Richard Curtis who continue to do so well those things that I hate to do. My editor Jared Kieling and his assistants, Ensley Eikenburg and Jesse Cohen, get a very special note of thanks for doing a terrific job in bringing this book to life. Additional help at St. Martin's Press was provided by senior publicity coordinator John Murphy and production editor Richard Klin.

Ms. Linda Price gets a tremendous amount of appreciation for her wisdom, talent and inspiration. Ms. Barbara Riefle, thank you for helping me with the original proposal. Janet Wiscombe, thanks for the encouragement and help with the title.

I would again like to thank my mom and the staff of the Franklin Square Public Library for fostering my love of books.

•

## AUTHOR'S NOTE

Because I have allowed these men to speak in their own voices, there may be some terms with which the reader is not familiar. The glossary at the end of the book explains these terms. An asterisk next to a name means that a pseudonym is being used to protect a person's identity.

# CONTENTS

**F**lames were blowing out the third-floor windows of the five story, dirty redbrick tenement. The rain kept the heavy black smoke close to the ground, obscuring the street lights and making the bystanders' eyes smart, while the acrid smell irritated their nasal passages. Gusts of cold wind temporarily lifted the smoke, briefly brightening the scene and freshening the air.

On the sidewalk opposite the building, elderly people shivered in night clothes. With them were children huddled against their mothers for warmth. Those old enough to be concerned wondered if they would have a home to go back to. Young men wearing high top sneakers and short jackets greeted each other with hand slaps. They half watched, concerned about looking cool in front of the girls or their "brothers."

Someone was yelling to the fire fighters dismounting from the first due ladder truck, "There's people trapped in the building!"

The fire fighters quickly took off their helmets to pull on the face masks of their breathing apparatus. With their helmets back in place, they looked like Darth Vader.

•

The captain led his two-man forcible entry team into the building. They ran up the stairs to the landing of the public hallway on the third floor. There the intensely heated black smoke that was filling the corridor forced them to crawl the rest of the way to the fire apartment. The only sound was their own accentuated breathing, which took on a mechanical quality when amplified by the masks.

The captain tried the door, but it wouldn't open. Forcefully through his mask he ordered, "Take the door."

His men understood the meaning of the muffled words.

With only the captain's light for illumination, the Irons man reached up and inserted the wedge on the end of his Halligan tool between the door and the frame. Then he said, "Hit it."

The probie drove the Halligan in with the back of a long-handled axe. The two firemen pulled on the other end of the Halligan. The door popped open, but it wouldn't move more than a few inches. "Christ, what do we have here?"

"Put your backs in it," ordered the captain. All three of them managed to force the door open enough for the captain to get his gloved hand and forearm inside. He touched the inside doorknob and found a taut wire connected to it. He ran his hand down the wire and encountered the neck of the dead man it was wrapped around.

When the battalion chief was told of their gruesome discovery he immediately radioed the dispatcher to send in the fire marshals.

New York is a city of fire. In 1989, greed, passion and carelessness produced over 101,000 fires, in which 246 men, women, and children died.

Whenever malicious fires are thought to have been set, fire marshals are called in to determine their cause and origin, and ultimately catch the arsonists. In 1989, New York's 239 marshals conducted over 10,140 investigations. Arson was deemed the cause of 5,362 of these cases, leading to the arrest of 132 women and 445 men. The New York City Bureau of Fire Investigation is the largest and busiest in the world.

Next to war, arson is humanity's costliest act of violence. Arson comes in many forms. Fires are started for revenge, for profit, for kicks, for recognition, out of fear, or to cover other crimes. As varied as the motives, so are the arsonists. They range from kids to little old ladies, from jilted lovers to Mafia wiseguys, from teenage crack dealers to lawyers and respectable businessmen. Arson is committed by individuals and by groups.

To apprehend these arsonists, the Bureau of Fire Investigation's marshals do everything from gathering physical evidence to mounting undercover and sting operations.

In carrying out their mission, fire marshals often become involved with crimes other than arson. In 1989 they conducted 1300 non-fire-related investigations.

The marshals' territory ranges from the poshest hotels of Manhattan to the ravaged streets of the ghettos in Brooklyn and the Bronx.

The first fire marshal in New York City was Alfred E. Baker. In the spring of 1854, while a reporter for the *New York Herald,* he was struck by the number of peculiar fires of doubtful origin, and he noticed that the authorities made no effort to find the cause of them. When he brought his findings to the chief engineer of the fire department, he was told that he could investigate these suspicious fires and bring the arsonists to justice. There was no pay, but the insurance companies raised a fund to remunerate him. One of Baker's first acts was the arrest of Charles A. Peverilly for attempting to set fire to a warehouse in lower Manhattan.

One year later, Baker obtained permission from Chief Carson and the Board of Engineers to wear the uniform of a fireman—red shirt, fire cap, and fire coat. The Board of Police Commissioners conferred upon him a sergeant's shield, on which was engraved Fire Marshal, New York. He was also given a regular salary.

•

In 1873, the Bureau of Fire Marshals was created in the Fire Department. Between May 21, 1873 and February 1, 1886 the Bureau made 142 arson arrests.

The name was later changed to the Bureau of Fire Investigation. By the 1960s, the bureau had a force of 73 men. On July 1, 1969 only fire fighters in the FDNY could become fire marshals.

By the middle of the 1970s, New York City was burning. The high point for fires came in 1976 when there were 153,263 of them. After the burning of the Bushwick section of Brooklyn in 1976, Mayor Beame ordered the fire department to appoint 100 additional marshals. The following year another 100 marshals were placed on the job. At its height, the Bureau had 405 men operating in three task forces. The number of fires and false alarms began to drop.

A few years ago, the shortsighted administrators began cutting back the number of marshals. Fires in New York City are again on the rise.

Today, in order to become a New York City fire marshal, an FDNY fire fighter must pass another demanding civil service test. After his appointment, the new marshal is sent for hundreds of hours of additional instruction at the fire academy on Randall's Island. He also takes the New York Police Department's criminal investigation course. Marshals are police officers. They carry guns. They make arrests.

Coming out of the firehouse culture where it is frowned upon to blow one's own horn, fire marshals tend to keep a low profile in front of the press. Few people know what their turbulent world is like.

What follows is based on in-depth interviews with some of the best fire marshals in the Bureau of Fire Investigation. Their compelling stories reveal their dangerous and intriguing around-the-clock battle to prevent the tragedies caused by arson and other crimes.

•

•

# FIRE MARSHAL

# JAMES McSWIGIN

\ was a "dirty-faced" fire fighter in the South Bronx. The people there became very attuned to the fact that if they needed help for anything, they could call the fire department. They knew if they pulled that box on the corner, they were going to have somebody there—a presence of authority—to help them, in a matter of three to four minutes. If they called the Police Department or anybody else, they might have to wait up to an hour. We used to get called for water leaks, stabbings, rapes, shootings, and robberies. You name it.

In my years in 27 Truck I delivered two children in the South Bronx. Both of them were named after me, but I don't know whatever happened to them. I also saw many guys get hurt. And I've seen a couple of good friends die in fires down there.

I think the death of Fire Fighter Dominick Rosato hurt the most, maybe because I was with him during his last moments. Ironically it was an

•

arson fire. A nickel and dime revenge fire between two Puerto Ricans that loved each other, but hated each other.

The date of the fire was May 21, 1969, box 3138, Daly and 180th Street. Funny how certain things stick in your mind for the rest of your life.

Twenty-seven Truck rolls in second due, at which time 38 Truck and 45 Engine had the fire knocked down and were throwing a mattress and box spring out into the street. The mattress was the origin of the fire. Twenty-seven and 88 Engine were told to "take up" by the 18th Battalion. Dominick was detailed that night to 38 Truck so he stayed.

As we are walking back to our rig we hear a gunshot and a lot of screaming. We turn and run back to the scene. I see Dominick lying faceup on the sidewalk. People are screaming, "Sniper! Sniper!"

In those years, the mutts were often shooting at the rigs and the fire fighters.

Total confusion and pandemonium. Panic was setting in. I was shitting in my pants wondering if the next shot was for me. Guys were diving under the rigs. The cops on the scene were taking cover and they had their guns drawn. I was looking down at Dominick and I didn't know what the hell was going on.

Frank Pellicci, our chauffeur, had some dude against the wall and was punching the shit out of him thinking he was the sniper. Captain Jack Miley was doing his best to get his men under cover. For those few moments nobody knew what the fuck was going to happen next.

I jumped down on Dominick and tried to talk to him. "Where are you hit? Where does it hurt? Get up, you bastard, we got to get out of here!"

Dominick looks at me with a blank stare and says, "Jim. Help me."

A kid from 45 Engine, Billy Cola, screams, "Turn him over. He's hit in the back." Billy is as scared as I am. Christ, I felt so vulnerable and alone.

•

We unsnap Dom's turnout coat and roll him over. Billy pulls up Dom's shirt and we look at his back. There is a small trickle of blood, and a hole no bigger than a pencil point right square in the middle of his back. I remember thinking, "Shit, he's center shot."

Billy and I start dragging Dom toward the chief's car, which is about 20 yards away. All of a sudden one of the cops starts opening up at the rooftop. Boom! Boom! Boom! Shooting at what he thinks is the sniper.

As we are dragging Dom to the car, I am saying an Act of Contrition, because I am going to die right here on 180th Street in this filthy goddamn ghetto. "Christ Almighty help us. Oh my God I am heartily sorry for having offended thee and I detest all my sins." It's amazing the great job the nuns of St. Frances de Chantal did on me. The Act of Contrition is right on my lips and I don't even remember thinking about saying it. But as I am saying the prayer I hear someone else saying the same words—it's Billy. We are both confessing our sins in fear of death. I guess he had the same nuns that I did.

We finally get Dom to the car, though Christ only knows how we got him in it. We have him propped up between us in the back seat. Bernie, the driver of the 18th Battalion, says, "We're heading to Fordham Hospital." Bernie has the pedal to the metal and he's taking the turns on two wheels.

We're telling Dom, "Everything is fine. You are going to make it."

We get on Crotona Park South and Dominick looks at me with a look I'll never forget, and suddenly he stops breathing. I'm numb. I'm blank and frozen. Billy starts yelling, "He's not breathing. Get on him. Get on him."

I push my mouth down onto Dom and start blowing air into him. Billy is pushing on his chest. Dominick is dead. But after five or six blows into his mouth, Dom responds by blowing back. I'm elated he is breathing again, but I have a strange taste in my mouth. It's blood. Not only do

I have his blood in my mouth, but his blood is all over my face. Dominick is alive, but with every exhalation he is blowing blood out of his mouth. "Oh fuck," I think, "he's lung shot."

Billy and I keep working on Dom all the way to the hospital. He kept coming and going. Dominick died two and a half hours later on the operating table. The kid never had a chance from the moment that slug entered his body. The .22 caliber bullet not only pierced his lung but also took off the top of his heart.

The only words that Dominick said from the time we got him into the car until he died at Fordham were, "Mommy, Mommy, Mommy," over and over again. Amazing, when the shit hits the fan, no matter how big and strong we think we are, it's Mommy we call for.

Here is the real kicker—it wasn't a sniper after all. The gun apparently fell out of the box spring that had been on fire and was picked up by another fireman and accidentally discharged.

After many crazy years in the ghetto I was tired. I thought, "There's got to be another way to fight fires than the fire department's Neanderthal way of beating doors down with an axe, and blasting the hell out of it with a 2½" hose line. I felt that as a fire marshal investigator with police power I could take away some of the mutts that were killing my friends and killing kids and adults in the South Bronx.

My first arrest is not a happy memory. I was a marshal only about three months and had done a number of small vacant-building fires, when I picked up an occupied-building fire with my partner Richie Zybryski.

We did the physical investigation and found out it was an incendiary fire, that is, the fire was deliberately set. It was a 6-story walk-up, multiple dwelling, about 150 feet by 150 feet, a big H type building over on Morris Avenue

•

and 175th Street in the west Bronx. It was of particular interest to me because I had fought fires there.

We started knocking on doors and introduced ourselves as fire marshals investigating a fire. As we questioned the people in the building, the pieces came together about two Hispanic boys who liked to play with matches and who'd been accused of making fires before. We finally had enough probable cause through sworn statements to arrest these two kids.

After arresting them we read them their constitutional rights. One boy's name was Sixto and the other boy's name was Pepito. The kids were 13 and 14 years old. Pepito, the older one, was a little slow.

I felt very good because we made our first arrest. I was doing what I was supposed to be doing. I was locking up the people who were burning the buildings down. In this case it happened to be two Spanish kids.

We took them to central booking and they made a full confession. Then we had to go to arraignment. But the arraignment wasn't going to be until the next day, so they released Pepito and Sixto in their families' recognizance.

The day after the arrest I'm still feeling very good and Richie is feeling very good. Then I got a call from the assistant district attorney. He told me that I didn't have to come in that particular day for the court proceedings. Naturally I inquired why, and he said, "Because Pepito went home after I released him. He went into the basement with a rope and hung himself out of remorse over the fire."

Pepito wasn't a malicious kid. He used fire as a way of striking out at misery in his life and it ended up with him killing himself.

If I had not arrested Pepito, he might still be alive today. It's a hard thing to rationalize—back and forth, "Did I do the right thing, or the wrong thing?"

I did the right thing. It just went bad. It went bad for Pepito.

•

5

At that particular point, I was as low as a crab's ass. I felt guilty that I had something to do with this young boy's death. So I had a big fight with myself at that particular time. I didn't know if I wanted to continue on as a fire marshal. I really thought about taking a demotion in rank and going right back to the firehouse where I had 15 years of helping people.

It left a very bad taste in my mouth, and there's still a bad taste in my mouth 10 years later. After this case I was very cautious with kids; as I find most fire marshals are. It's a bad situation to be in. We encounter a tremendous number of seven-, eight-, nine-year-old kids involved in fire. In New York you can actually lock kids up at seven years of age, under the criminal procedure law, for the crime of arson. Fire marshals are family men, and they really are hesitant to make an arrest of a kid at that age. But it's a bum trip because that kid out there is basically a time bomb.

Richie and I stayed partners for a couple of years and we were the top guys for arrests and convictions in the Bronx. Richie and I just clicked together and we were getting a tremendous number of collars. Keep in mind these were not arrests for drugs or anything else—just arson arrests. There wasn't a set of tours that went by that Richie and I didn't have somebody in handcuffs, parading them in front of the courts. We thought we were doing something productive for the city, for ourselves and for our brother fire fighters. We enjoyed it.

Most of the arrests we were making were for revenge fires. The guy who makes this fire is trying to get even with somebody else for an imagined or real travesty against himself or his family. I'll give you an example: The guy is pissed off that his girlfriend is seeing another man. This type of fire setter is probably one of the most dangerous because he's usually enraged when he's starting the fire. But he's usually the easiest to catch. In his rage, he'll try

•

to burn her out, or burn the new boyfriend, with a Molotov cocktail or by squeezing flammable liquids under the door, and he won't really care about who sees him or about anything else. Consequently his crime of arson isn't surreptitious. Revenge is probably the second most frequent cause of arson fires in the City of New York and in the nation today.

The "landlord fire" or the profit-oriented fires are about fourth on the list. Number one is kids. Juveniles set more fires than anybody else. They probably start 50 percent of the deliberately set fires in this country.

Women also have no hesitancy whatsoever in starting fires. I have locked up quite a few of them. Most women's fires are for one or two reasons: either spite-revenge, where they're pissed off at their husband or their boyfriend and they were going to get even with them, or welfare fraud.

Here in New York, the latter is a profit-motivated fire. Because in New York, if you have a fire and you're on welfare, you would go to the top of the list and get the next best available place. So, to beat the welfare system, they would start fires in their apartments to improve their surroundings by getting their family moved into another neighborhood.

Most gals, though, make fires to get even with husbands or boyfriends. You'd walk into a fire scene and it will be the kitchen that's burned. In your physical examination you'd look around the kitchen and on the stove you'd find a mountain of metal zippers, metal clips and burnt cloth. You knew by looking at that stove just where the fire started and the reason for the fire, without ever talking to her. It doesn't take too long to figure out she's very pissed off at her man. She took all his clothes and piled them on top of the stove and turned it on.

We locked up a number of women, because they broke very easily. If you got a few pieces of information by your

•

physical examination and you were slick enough you could approach a woman and make her confess very, very quickly.

A man will hard ass you. "Screw you. I'm not talking to you."

A woman will break down when confronted with authority much faster than a man. I don't know the reason for this. Maybe women are the fairer sex. Because when they know they've done something bad, they wear their guilt all over their face.

We had one gal up on Carpenter Avenue in the Bronx who was a particular pleasure to arrest, because she had two children who looked as if they'd been knocked around quite a bit. When we started the investigation, her name and her face struck me as familiar for some reason. Then I remembered that a few years back, one of the guys had this particular woman in for questioning. She had burned down five different apartments while on welfare. She started off in Harlem, and she burned her apartments all the way up into the north central Bronx.

We nailed her because we had a couple of eyewitnesses who saw her running from the building. She had used flammable liquids, and we had her fingerprints on the containers.

God only knows what the value of the five different apartments was, and God only knows what value to put on the injuries to the fire fighters who responded to the woman's five fires. I don't really think that she spent any time in jail. I had gone to the grand jury and got the indictment, but we never went to trial. We got into the plea bargaining situation that we have here in New York. Her lawyer made a deal for her to plead guilty to a misdemeanor rather than be tried for a felony. She spent a couple of nights in the slammer and the court put her on probation. However, on the good side, I've never heard of her making another fire.

We received a call for a fire on the third floor of a building

•

on the Grand Concourse in the Bronx. Along the hallway on the third floor there were 10 areas where you could see where little piles of rubbish were placed and set on fire. They looked like what we call in the business "female fires."

In interviewing people, a gal by the name of Marjorie was singled out as the one always seen in the area, just prior to these little fires breaking out.

When we got into Marjorie's apartment, in the living room alone I could see that about 15 fires had been started. There were burn holes all over. She had a four-room apartment and there wasn't a piece of furniture that didn't have scorch marks or holes burned in them where she set these little fires and then extinguished them. But they could still cause structural damage.

We had enough probable cause at least to bring her in. We asked her if she would like to come voluntarily to the police station, which fire marshals use as a base of operations when making arrests. Well, Marjorie confessed to hundreds of small fires. She had been doing it for years. But the more we talked to Marjorie, the more we realized that we were not dealing with a woman who was mentally stable. Consequently, when we went to court we suggested that she be examined by a psychiatrist. As it turned out, Marjorie was admitted into a mental health facility. I think she spent about six or eight months as an inpatient, and today she's still being treated as an outpatient.

Her fire setting behavior was broken. The arrest that we had made did not hurt the woman. In this particular case it helped her and it certainly helped the fire fighters and the people in that building who were plagued with these fires. Under the right set of circumstances you could have had a holocaust on your hands.

Some fire investigators might classify her as a female pyro. A female pyro, according to the books, usually starts fires on the floor that she lives. They will be small fires, so that after she starts it she can run quickly back into the

•

safety of her apartment, close the door and look through the peephole to watch the confusion that she has caused. A female pyro living on the first floor usually will not go to the roof to start a fire.

The local engine company gave us a call. "Since you guys locked that gal up we haven't been back in that building." Before, they were there every single night. It's a good feeling when you can do something like that.

I know one kid—I shouldn't call him kid—he was about 21 years of age, and he can be classified as a male pyro. Tony is his name. He was involved in many fires in Queens. Tony was a "square badge," a minimum-wage guy who operates as a uniformed, private security guard for places like bakeries, trucking concerns, and residential buildings.

Tony can also be classified as a vanity-fire setter. Because not only is he the guy that makes the fires, but he also discovers and extinguishes them in order to receive the praise and accolades of his peers or his superiors. A number of our marshals had been on to Tony on previous cases, but he was playing a game with us.

I was working with a guy named Gene Skelley at the time we were called in. In all there were about six or eight marshals who had investigated fires that Tony had been involved in. One particular guy, George Scott Clanton, wanted to get Tony in a bad way because, at the last three fires he felt Tony was a suspect in, several fire fighters got burned. Tony's fires seemed to be getting bigger and bigger. But Tony was cute. There was just no way they could nail him. There was not one eyewitness. So, Scotty Clanton came to Gene and I and said, "Listen, I need a fresh face on this case. I've been hounding this guy. But I need a different approach. I don't know how to get to this guy. The kid's got heavy duty problems. He's making fires. He's going to kill somebody."

Gene and I said, "Okay, we'll ask Tony if he'll come

•

and meet with us down in headquarters," which at that time was at 110 Church Street in Manhattan.

Tony, figuring this whole thing was a game, played it out. He said, "Sure, I'd like to come in. You fire marshals are accusing me of making fires and I'd like to get this thing straightened out."

We sat down and read him his constitutional rights. We told him that any time he wanted to leave during the interview he could leave. Then I used a completely different approach. Instead of the normal investigative questions you would ask, I got into his childhood and on one or two questions I noticed I really hit a sensitive chord. I could read his body language and there was something there; as I dug deeper Tony started to open up. But I still did not know exactly what vehicle I was going to use to get Tony to confess to what we suspected were about six or eight fires. You look for the right tack to take.

What came out of that interview was that Tony at 21 years of age never had sexual relations with a woman. Tony was deeply involved in the homosexual scene. The gay leather bars, like the Ramrod and the other joints on the West Side. Tony had been sodomized by his stepfather from age 10 or 11 up until the time he was 16 or 17.

I went on with the interview. I wanted to find out where was the chink in his armor. I found it with the homosexual scene. After about an hour and a half, I looked at Gene Skelley, but I said to Tony, "Do you know why Fire Marshal Skelley and I have been brought into this case?"

I hadn't said anything about fires yet.

"Fire Marshal Skelley and I have been brought into this case because we were both heavily involved in the gay scene, but we straightened our lives out and now we're both married men, and we're happy. We're here as the homosexual squad to help you."

Well, Skelley almost shits. He turned red as a beet and his eyes were the size of silver dollars. Gene has a great

•

11

sense of humor, but right now he's pissed. He's sitting behind Tony and he's sticking his tongue out at me. I can tell by his eyes that Gene would like to strangle me. I'm trying to keep a straight face. But as I'm going through with this I see that Tony is buying it. Now a fire marshal or a police officer can use a certain amount of trickery. This is trickery, no two ways about it. Gene and I are not of that persuasion, but it was the way to go.

I said, "Tony, I'm not even going to discuss fires with you. Fire Marshal Skelley and I are here to help you. In all probability you will be arrested in a week or two; but we have to talk to the chief fire marshal and find out how many charges of arson we're going with." Although we really had nothing to charge Tony with, we went through this big fanfare of slamming the file folder closed and saying, "Tony, you're free to go now. Get the hell out of here! The next time I see you you'll probably be under arrest."

I stood up and Gene Skelley stood up. Skelley was still making faces at me. He's very upset with me for using that tack.

Tony looked at us and he said, "Well, fellows, sit down. Maybe I'd like to talk to you."

At that moment I knew that we were going to nail him. He was going to tell us about the fires. Gene and I sat back down again. The funny thing with a pyro, or any multiple fire setter, is that the hardest thing for them to do is to tell you about the circumstances of the first fire; once they do, they will tell you about any other fire they're involved in that you want to know about. But the first one is the bitch.

Tony starts off by saying, "You know, Jim, I can empathize with you guys. I see you're here to help me. You're not trying to hurt me."

I said, "Well, that's it, Tony."

I'll tell you, I wanted this kid in. This kid was a goddamn time bomb.

•

Then very slowly he said, "Well you know. . . . sometimes when I'm in the apartment house I get cold. . . . and I would shake. . . . and when I made the fire. . . . I would feel warm. . . . and I would get better."

He seemed to like the confusion and the game that this whole thing created, with the fire department coming in, the lights, the sirens, the people running around, and him directing them to get out of the place.

As a fire investigator you ask questions *around* what you're getting at. For example, you will make statements like, "Well, to me that fire looked like there was paper involved. Was there paper any place around the apartment?"

You get an answer, "Well, yeah. There was a lot of cardboard and paper in the bathroom."

"Then you probably took the paper out of the bathroom and put it around?"

In other words, you're going around the horn to get him to admit to the ignition. You don't want to alienate him by direct questions like, "Did you start that fire?" Because if you back him up he will pull his shell around him again. Sometimes it doesn't work, but in this particular case we felt that we should keep asking questions around the actual fire.

When he had painted a complete picture of where he found the paper, where he put the paper, where he found the matches, then we get him to tell us what he did after the fire was started. Finally, we arrived at the question: "What did you start it with? Was it a lighter, a match, or a candle?"

"No, I used a match."

Now we have not only probable cause, for which you can make an arrest right there, but he has actually told us he's taken the match and made this fire. Now the floodgates are open. Tony told us about eight other fires.

Then I said, "Tony, we're going to bring you out to the

•

district attorney in Queens, he's a good friend of ours and I think he's also a former homosexual. I would like you to tell the district attorney the same thing you told us."

What we're trying to do is build an ironclad case. We went out to Queens. The assistant district attorney (ADA) brought in a tape recorder, turned it on and explained who was in the room and read Tony his constitutional rights. The ADA said, "He's 21 years of age. He feels that he wants to get help for himself."

I'll tell you I wanted to get help for him. But I also wanted to get the SOB off the street. Two security guards who worked with him were severely burned in one of Tony's fires. So it gets back to, yes, I'm very concerned about his sad career, but I also owe the people of New York and I owe the fire fighters. I have to do something about this guy. He is a menace.

Well, Tony starts talking to the assistant district attorney, and Tony gives him 31 fires. He knows approximately when they were and where they started.

Scotty Clanton makes the arrest and Tony goes into the system. Scotty now has to check out all these other fires. Sure enough, on or about the dates that Tony gave, and at those addresses, there were fire department responses for fires.

My function in that particular case as a fire marshal was completed. I was the fresh face. Although Skelley and I still laugh about it, I found the vehicle to get Tony to admit to these fires. Without his confession, there was not enough evidence to arrest him, much less get a conviction in court.

I later found out he was only charged with six of the fires. He was remanded to some religious leader out in Queens and he spent some time with him in the rectory. The religious leader kind of took Tony under his wing, but that's another story that I would rather not get into.

I just heard some sad news that about a year ago that guy who was his guardian had passed away, and Tony's

•

out again. Tony's upset. Tony's making fires again. So I think the day is coming where I'll probably be getting a call from some fire marshal who knows his story and I'll go back in and see Tony again.

At a suspicious fire I often stand back and watch for the guy who helps the fire fighters. He's the very first guy I would talk to, because he may very well be a vanity-fire setter. You see this a lot with the square badge, because once their boss says what a wonderful guy they are and how he wants to give them a citation, they become heroes.

I've had a few of these guys. I locked up a guy at the Trinity School, which is a very prestigious prep school over on the West Side of Manhattan, in the nineties. They had a series of fires. One of the questions a fire investigator asks is, "Who discovered the fire?" That's an important piece of information.

"Tom discovered the fire."

"Who's Tom*?"

"Tom is the security guard."

"He was praised by his company, and the school gave him a citation."

"Oh, have you had any other fires?"

"Oh yeah, we've had 10 small fires in the past couple of months."

"Who discovered the last fire?"

"Oh, Tom did."

As we asked more questions the administrators of the school started to see where we were going. Tom was at most of the fires and he had discovered most of them. At some fires he even stretched the hose line. At other fires he warned the whole building and got everybody out. We finally built a nice case against Tom and locked him up.

In these types of investigations, although they sometimes can get a little lengthy, you can often successfully conclude them.

Investigations dealing with wiseguys, merchants, or landlords, however, are extremely difficult. They require

•

a tremendous amount of work: Many, many interviews, and much digging, not only in the rubble but also in the business records.

I had two guys out in one of the boroughs who were tied to organized crime, for lack of a better word. They were involved in much shit. They had a gas station which wasn't doing well, so they blew it up. Though the tanks didn't go, they had a great many tires in the place, and they got a good fire going with flammable liquids. The battalion chief and his men described the flames as having many sparkles in them, along with a strong odor of gasoline.

I'm sure they used some type of timed, incendiary device to trigger the actual ignition of the fire. But I could not find the device in the rubble, or the exact ignition point. Instead of a point of origin, I found an area of origin, meaning that the whole first floor of this place was saturated with gasoline.

I canvassed the neighborhood and I came up with some kids who worked for these two guys. They were mostly poor black kids, but they gave us a lot of information.

I used our power of taking sworn testimony. A fire marshal in the City of New York has not only police powers but he can go to a witness and the witness can write out a statement that he saw Joe make the fire. The fire marshal then asks the witness, "Do you swear that the statement you gave me is the truth?"

If the individual says, "Yes, I swear" the fire marshal can say "Okay, now sign it."

Once that guy signs the statement, it's a sworn oath of affirmation; it can't be changed in court. If the guy later on decides he is going to recant, then under the law he has lied to the fire marshal and I can arrest him for perjury. It's a very effective tool in fire investigation. I wish our police officers had the same powers. Once you get that sworn statement that a witness has seen this guy starting a fire, it's an easier investigation. Of course, you still have

•

to go through the whole investigation to decide if what they're telling you fits the facts of where or how the fire started. In other words, if they tell you, "Yes, I saw him start the fire in the bedroom," but your examination proves that the fire started in the living room, then their statement really isn't worth anything.

In all I got about 18 witnesses stating the facts of this particular fire. For example, on the evening of the fire the owner and his cohort asked the kids to tear the sign down from the top of the building, because they said, "We're not going to be in business anymore. Tomorrow we're going to be out of business." Approximately an hour before the fire, two kids were sent to the gas station up the block to buy 20 gallons of gasoline, and they carried it back in 5-gallon cans and brought the cans into the premises. Because this guy was in financial trouble, the power company had shut off the electricity to his building, and he couldn't run his pumps to get his own gasoline.

Kids working at another gas station heard the explosion, and saw Angelo*, the owner, and his cohort racing his Cadillac across the gas station's tarmac to beat the light and get the hell out of there.

I got statements from people in the local hardware store that Angelo had purchased 9-volt batteries the day before the fire.

I had a barmaid who heard them laughing and joking about how the business was going to be gone tomorrow. How it was going up in smoke.

We want to get back into the fire building to complete our physical examination. The owner lets us in, but he's trying to keep us out of certain areas.

In the meantime, my partner Richie gets on the telephone. He called up to find out if there are any warrants on the two guys we're interviewing. Well, one of them has an outstanding warrant on a gun charge. My partner notified the cops, which was probably a mistake at the time.

•

The cops arrive at the scene and they collar Angelo on this gun charge. It figures when you don't want the cops right away they show up in less than 15 minutes. When they arrest him, however, the gates come down and the whole place was locked up. Angelo, however, makes bail right away.

A month after the fire, I went to the assistant district attorney with what I felt was probable cause to make the arrest. He suggested we approach the two individuals and threaten them with the RICO Act to get one of them to fess up. I was completely against it, because you're not dealing with the mutt in the street, you're dealing with guys who know the game. They know the system. But the assistant district attorney insisted that we go try this first.

The district attorney's office is fine, and they do a great job with their limited resources, but with some of the ADAs you give them a case on a silver platter they'll still want you to do every goddamn conceivable thing to give them the easiest case possible in court. We had a great case. Two months go by and Angelo and friend get themselves a lawyer. The lawyer's name is Mario*. Mario is strictly known as an organized crime lawyer. He's good, there's no two ways about it. Mario is also a former assistant district attorney in that particular county. He knows all the ADAs. They're all very good friends.

In 1978 I parade in front of the assistant district attorney with this case. The district attorney says, "Okay, we're going to get working on this."

Fine, we go back and I continue on the fire investigation. We have a grand jury presentation coming up, but I figure we're doing good. They're wiseguys, and we are going to nail them by getting an indictment. Well, a month goes by, three months go by, six months go by, and I'm calling the assistant district attorney to ask, "What's going on?"

"Oh, we're working on it."

A year goes by. I have this case sitting out there. It's a lulu, 18 witnesses, a beautiful case. Then we get a call from

•

the assistant district attorney, "We got to ask you something."

"Yeah, what's that?"

"You have copies of all that case material you brought out to us? We seem to have lost it."

"I've got copies of everything. I'll be out to your office tomorrow."

I personally think there's a rat in the woodpile; something went on between someone in that particular district attorney's office and their former assistant district attorney, who is now an attorney for organized crime figures. This is a money case.

A year and a half went by before it ever got in front of a grand jury. I had to reach out to Washington to get one of the witnesses back, and we had to go out to Maryland to get another one back. But they all came and testified before the grand jury. The grand jury gave us a true bill, which means an indictment and gives us a warrant to arrest these two guys.

In addition to arson I wanted to charge Angelo with mail fraud. Because if he committed a crime and uses the U.S. mail to send in this insurance claim, or uses the mail to facilitate or perpetuate this fraud in any manner, it's a federal offense.

The assistant district attorney said, "No, we'll get this case going," and shortly after that we went to trial.

I had to stay outside the courtroom until I was called to go on the witness stand. When I got on the stand there were two defense attorneys, one for each of the suspects that I arrested. Well, I'm telling you they did some number on me in court.

The average person can only offer into evidence what they actually saw or heard. It is what we call direct evidence. A marshal, however, is recognized by the court as an expert in the field of fire investigation and can offer into evidence his opinion. That's good in many ways, but it also opens the door for the defense attorney to come after you and ask you questions that you'd better know the answers to or he'll make you look like a horse's ass on the witness stand.

We found that the gas meter was beaten off the wall, but there was no gas leak. Although the defense tried to prove that the fire started from the natural gas, I had statements from Con Edison that the gas was shut off in the street a week prior to the fire.

The funny thing was, while we were at the scene investigating the fire we knew we had a flammable liquid fire, and that this was an insurance fraud fire.

In the court the defense attorney broke my balls for three hours over my physical examination. "How come you didn't complete your investigation?" The defense attorney knew full well the reason I didn't finish my interview with his client was because the cops locked him up. Had I said this in court it would have gotten the case thrown right out, because it would have prejudiced the jury.

Testifying is horrible. Though you're schooled to a certain degree, when you get up there you have 12 jurors, a judge and all the other people in the courtroom looking at you. For the average guy who's trying to do a decent job, trying to do what you're being paid to do, what you're sworn to do, it is hell-raising because the defense attorneys are trying to make you look like a bastard. How dare you make these accusations against my client? They abuse the shit out of you on the stand. It gets to be a game between you and the attorney. But stop and think about it. How is a guy who spent most of his life as a fire fighter supposed to do verbal combat with a high priced, educated lawyer? I mean these guys are sharp individuals. Fortunately, I happened to do well on the witness stand—according to the district attorney—although there were times I felt as if I just wanted to crawl inside myself and hide. But the defense attorney was doing exactly what he's paid to do.

When the trial ended, the two guys were found innocent. I think there was a deal. It seemed that someone

•

waited a very long time to get a particular judge, because there was no reason why this could not have gone to court a lot sooner. Once you get a true bill you're supposed to make the arrest and start the trial. Well, we waited a year before we came to trial. I wasn't too happy with the judge. I couldn't say if he was or he wasn't corrupt, but something stunk.

When the jury came in with a "not guilty" verdict I was down in the dumps. I figured I put a lot of work into this case, and these guys definitely made the fire. There were no two ways about it. How does a prosecutor's office lose a case when the documents alone were about 12 inches high? That's just the statements of the witnesses and the business records which showed the motive.

This one assistant district attorney, Abe*, and I stood outside as the jurors came out. We stopped and asked a few of them, "What did we do wrong in the case?" I was using it as a learning experience. The three jurors that we managed to talk to said, "Oh, fire marshal, there was no doubt that he made the fire."

"Why wasn't he convicted?"

"Well, you did such a good job proving how he had spent so much money on cars and how he was so deeply in the hole, that we felt sorry for him." Then the one guy jumped in and said, "Listen. Really, the only one affected by this is the insurance company. Nobody was hurt."

I said, "What about the fire fighters that risked their lives to put that fire out? What about all the toxic shit that they breathed in? What happens when your insurance rates go up because the insurance company might have to pay up on his one hundred and fifty thousand dollar-claim for the business?"

It's a sad commentary. They never thought of that. And a good defense attorney will make sure when he picks his jurors that he'll pick business-oriented people, people who can empathize with this particular mutt's plight. Guys in small businesses know how hard it is with the insurance

•

and the bills and everything else, and how easy it is to get in a hole.

There's a good side to the story. This mutt Angelo decides that he's going to sue the insurance company for the $150,000 because for over two years they haven't paid him off. Why should they pay him off, since I have him charged with arson in the second degree and a number of other things?

Well, there are two courts. There's criminal and there's civil court. Now in civil court, the game changes; I don't have to be on the defensive. In civil court I can say, "He was locked up while I was on the scene and that's why I couldn't complete my physical examination." I can also say, "That son of a bitch started the fire." I can say many things and introduce a lot of evidence that I can't in criminal court.

Now when Angelo wants to sue the insurance company for $150,000, guess who goes to court again? Yours truly. But the insurance company decided to settle with him. They gave him about $18,000 on his $150,000 claim, which was probably not enough to cover Mario the lawyer. So I said, "All right, it's my turn now. No hard feelings, fucko, but I just took one hundred and fifty thousand dollars out of your pocket." I didn't get him criminally, but I stuck him where it hurts.

It was truly amazing when you got into investigations in the Bronx, because you'd find that the landlord was so and so's brother. Congressman who?

I remember two of our marshals, Tony Lopez and Wesley Powell, locked up a Brooklyn judge's brother for arson, on the same day that the mayor was swearing in the judge at City Hall. The brother was a mutt. But the heat that came down on these two marshals was intense. A number of politicians came to this mutt's defense. "How unscrupulous these fire marshals are to dare attack the integrity of judge so-and-so's brother, who is a marvelous individual."

•

What bullshit. He's a mutt who made fires. He burned a couple of buildings. I know Tony and Wes were really raked over the coals. These politicians wanted to hang them by the balls. They wanted to lock them up. "How dare they. . . . This was a travesty" these scum suckers were screaming on their microphones. "What dastardly deeds these police and fire marshals are doing."

These two guys were worried for a long time. But there was a collar made and a true bill handed down by the grand jury, though the case washed. That's all part of the game.

One thing I'm guilty of is shooting my mouth off. But what I'd like to say, after 25 years in this department and being involved in many of the investigations of fires in the South Bronx, is that the burning of the Bronx in the sixties and seventies was nothing but an insidious plot by powers that be to either gentrify or commercialize the place. It was done with at least tacit approval of some of our politicians, who were heavily invested in the burning and rebuilding of the South Bronx. At one time, though, I thought that was complete nonsense.

When I was a fire fighter in 27 Truck, we caught a fire on Fulton Avenue. After we had spent about six hours at this fire we dragged our asses down to the street and helped 46 Engine pick up their hose as the sun was coming up. It was early in the morning, 5:30, or 6:00 A.M. There was an old black gentleman sitting there and he asked me, "Why do you firemen bother putting that fire out?"

I looked at this guy as if he was a screwball, and I said to myself, "What am I dealing with here? Am I dealing with some wino?" The answer to that was no. He was a very distinguished-looking gentleman: Suit and a cane, an elderly black man probably in his late sixties, early seventies, but very well kept. So I said to him, "Good morning, pop. What do you mean?"

"Why do you put that fire out?"

"Well, we're here as fire fighters. We're here to protect

•

lives and property of the people of the City of New York."

"Mr. Fireman, don't you know this is all supposed to burn?" And I looked at this guy and I said, "No." He might be a very nice-looking gentleman, but I'm dealing with a fool who says that this is all supposed to burn. Do you want to know something? I was the fool, because he was right. The South Bronx was supposed to burn, and burn it did.

The powers that be were faced with an area primed for industrialization, yet was filled with block after block of six-story walk-up tenement buildings, filled with blacks and Hispanics.

During the heyday of the sixties and seventies, a strange thing was going on. Construction companies were putting in new sewer, water, and electrical lines where the residents were shooting at and often killing each other, and where occupied and partially occupied tenements were burning every night.

We fire fighters asked, "What the hell is the matter with these people? Why would they be putting in all these utilities for an area that's burning to the ground?"

The federal government during the Johnson and Carter years made available a tremendous amount of no interest or low interest loans to rehab tenement buildings. Well, with millions of federal dollars coming in, it was like ringing the dinner bell for the landlords and the politicians.

Bill Moyers, in his journal on the Bronx, talked about the unscrupulous landlords who got tied up with unscrupulous insurance agents to make a criminal arson ring. They buy, or already own a building, and they insure it. They hire themselves a torch, and everybody gets a nice piece of the pie.

When you have a building where the top-floor apartment burned, it is usually an indication to us that the landlord is involved. Number one, the top floor is away from everybody in the building, so you get an apartment on the top floor and let it become vacant. Usually it's in

•

the back of the building. The reason for the back of the building is because people on the street can't see what's going on. Number two, in order to get the most insurance proceeds from a fire you have to remove the roof from the building. If you start a fire on the first floor there's a very good chance the roof is not going to burn off, but if you start the fire on the top floor, rear, it burns the roof off the building. This leaves the place exposed to the elements: Rain, snow, and wind cause further damage.

If you want to vacate that building, the fire fighters, my brothers, will come in with their hoses and pump literally thousands of gallons of water into that top floor to extinguish the fire. The water's got to go someplace, so it goes to the lower floors, and you have water damage. So with one fire in the top-floor rear apartment, that landlord can successfully get all his $80 and $100 a month rent-controlled tenants out. He also gets the maximum money from his insurance company.

Now, he has a vacant building for which he can apply to the federal government to get a no interest or low interest loan for rehabilitation. He'll put the roof back on it and make some attempt to fix the building up, and he'll insure it again.

Then geez, another fire and the whole goddamn place burns to the ground again. He double dips—he gets the insurance money again, he skates on the federal money since he has no tenants, because the building's ruined; he can't meet the payments on the federal loan. But most of that money is still in his pocket.

He walks away from the building and refuses to pay the taxes on it. The City comes in, condemns the place and knocks the building down. Quite often, the guy who had the property before bids on it when the land is auctioned by the City for back taxes, for a dime on a dollar.

Isn't it wonderful the way this works: They made hundreds of thousands of dollars on that lousy 35 by 75-foot 6-story building, while my brothers are going in there

•

and getting killed and maimed; all the while watching the blacks and Hispanics being used.

After all the burning was finished, a marvelous thing took place: Out of the rubble and ash rose industrialization. Amazingly, all they had to do was connect up the utilities put in the ground 10 years earlier. Some people are so *lucky* to have had the foresight to invest in those properties.

Back then, they only had 60 marshals, which they expanded to 100—and that was the City's major commitment to stop the arson. It was only window dressing, because they actually reduced the number of companies and the manpower of the fire department during the height of the burning. The funny thing is most fire fighters, and many marshals, feel that the effort of the Red Cap marshals was the real reason that fires are down in New York. Well, I can't take that away from them.

A Red Cap unit is a group of specially designated guys that use an overt approach. They're in a particular area, usually with a command trailer. They're highly visible. They drive around in red-and-white marked cars and wear red baseball caps with a fire marshal patch on them. They respond to every fire within their designated area in a given period of time. We've found by just simply investigating every fire in a particular neighborhood that the number of fires drastically goes down. The false alarms go down, while the arson arrests go way up. It is proven that it can work. Which makes me think, if that's the case, why aren't we investigating every fire in the City of New York and why don't we have a thousand marshals, rather than no Red Cap program at all?

I first came on the marshals with 50 other guys and we were heavily involved in the Red Cap program. We considered ourselves shock troops. The Red Cap program was quite a bit different than it is now. It was more physical.

We'd roll up on a building that had a number of incendiary fires, and two teams of guys would sweep the place. We'd work from the first floor right up to the roof. Quite

•

often, we would catch people in there, whether they were making fires, stripping the building, or selling drugs. We would bring them up to the roof to atone for their transgressions: Two burly marshals would make them recite the fire marshal's pledge that they would never, ever, set foot in a vacant building again.

During one of these sweeps we hit a building on 178th Street, in the Belmont section of the Bronx. It was a vacant building. But our stats said that over the past couple of weeks this building had a good dozen fires in it. So we would stake it out from time to time, or if we were away on another job we would just sweep back in and go through every room in the building.

One time, we see this dude come tiptoeing down the block. And as he comes down the block he looks up the block, and he looks down the block, then he goes into the building. Right away we figure we have ourselves a winner here. We decide to give him about five minutes before sweeping the building. We don't want him to get the goddamn place roaring. We figured we'd nail him in the incipient stages of fire. We hit the first floor and we're sneaking around with our pistols drawn, poking into every room. We get up to about the second floor. You have to realize it's late, about 1:00 in the morning, in a completely dark building, and the only light you have is your flashlight. As you're looking through these rooms the hair is up on the back of your neck. You know he is in the goddamn building and you know he is not supposed to be in here. But you don't know what he has on him.

I came into this one apartment, and I peeked around the corner. No more than two feet away from my face is this face looking right at me. I must have gone two feet into the air, screaming, "Ahhhh!"

He also screamed, "Ahhhh!"

I yell at him, "Police! Put your hands up!" But he had his drawers down. Now this poor guy is standing with his hands up in the air and he goes to grab his pants. I

•

scream, "Don't move motherfucka! Don't move! Police! Police!" So now I've got him standing there and the pants are down around his ankles, his hands are up in the air and I've got the flashlight on him. I said, "Son of a bitch, what are you doing in this building?"

He said, "I just come in to do a doo-doo."

I said, "Doo-doo my prick. I should lock you up. I almost shot you. You son of a bitch."

Now the pressure is off. I got him with his hands up because I don't know what he has in his pants, although at this point, I am not going to look in his pants to see if he has a knife or a gun. He can have anything he wants. I called the other squads in from the rest of the building and we conducted an interview. We went through all kinds of questions, until a couple of marshals started to gag with the smell.

Finally we decided to let him go, but not until he made a fire marshal's pledge to us that he would never, ever come back into that building and that the fire marshals were the most wonderful people. They also had to sing this little song that had a line in it that said, "The fire marshal is my friend." Then we let him leave, but he wasn't allowed to reach into his pants; so he shuffles out into the hallway, with his drawers around his ankles, down a flight of stairs, out into the street. We had quite a laugh over it.

We found that we were shock troops, for whenever we went into an area we really put the lid down on not only the fires but crime in general. There were so many marked fire marshals' cars in the area and we were running into all kinds of things.

Many times, though, your partner is in one building and you're in another. You're out there by yourself; often a white honky face dealing in the bowels of the world. You really depend on your own wiles to survive in the street. I was considered to be a good street marshal because of what I learned as a fire fighter in the South Bronx. After

10 or 15 seconds on a block I could tell you who the drug dealers on the corner were or what some of the other suspicious-looking characters were up to.

We used to introduce ourselves to the drug dealers on the corners, because we weren't too interested in drugs. That's a police matter, we figured, let the cops take care of them. But we used to get some of our greatest information on who was burning down the buildings from the drug dealers, because they didn't want us in their face. Their game was selling drugs, and they wanted this heat out of their particular little bailiwick, so to speak. They didn't want these Red Caps running around, kicking ass, and causing general dismay in the orderly sequence of business that they had going in the South Bronx. We made a lot of arrests.

At that time the supervising fire marshal was Tom Sweetman. Tom is now a deputy chief.

Richie Zybryski and I made arrests like gangbusters. Richie had come out of a busy firehouse in Brooklyn, and he had that ghetto fire fighter's zeal. We felt that this was our turn to put the happy match brigade on the run. But whenever we got collars we would bring a little levity into it.

We made the arsonist call up our supervisor and ask for "the chief fire marshal of the city, state, and United States—Mr. Sweetman." They had to apologize to Mr. Sweetman for the fire they made.

Well, the first time that Tom got a call he didn't know what was going on. I took the phone and said, "Tom, we have an arsonist here. He's a confessed arsonist and he's throwing himself on the mercy of the court and he wants to ask for your forgiveness for making this fire."

We made it a ritual. About 14 or 15 times where we had collars in which the guy or the gal fessed up to the fire we made them call Tom Sweetman, the chief fire marshal of the United States, and apologize. Sweetman had gotten into the game and he would say, "Very good, I accept

•

your apology. Now what I would like you to do is cooperate fully with Fire Marshals McSwigin and Zybryski."

The guys would come into the office and say, "Well, Sweetman got a call last night," or "another guy fessed up." It was a hard job at the time, but we had a lot of fun with the crazy things that happened in those years.

I recall one occasion when we were asked to come in to assist with an excavation in Harlem. There was a clothing store on 116th Street, and early one morning its show windows fractured into thousands of little shards as the building literally blew up. The battalion chief made the fire "suspicious" and the marshals were notified.

Wesley Powell and his partner began asking people in the neighborhood about what they knew. It turned out that a guy named Pedro, who worked in the store, happened to live around the corner. Pedro was seen in the area just before the explosion, but not since then. The next day, Powell visited Pedro's apartment. His wife said that he had been missing for two days.

It also came to light that Pedro's friend made a sudden trip to Puerto Rico by himself.

Now Powell was highly suspicious, and he felt that Pedro might still be in the rubble of the building, even though the secondary search performed by the fire fighters after the flames were darkened down had been negative.

Some time had passed since the investigation was started, during which the remaining part of the building had been torn down and the area filled in with dirt. Arrangements were made by our office to secure a search warrant and bring in a backhoe to dig this lot up. One of our bosses, Chief Mike DiMarco, called us in.

We put up barricades, and the backhoe went into operation. A crowd of onlookers and hecklers had arranged themselves around the barriers. The digging went on for hours. Periodically the cops would drive by and harass us.

The state police were enlisted in this operation because they have a dog trained to sniff out bodies. The dog looked

•

like a mangy mutt. It moped around the lot, doing what dogs usually do. Whenever a new pile of dirt was deposited by the hoe the dog was directed to sniff the pile. At one point its trainer produced a jar, opened the lid and lets the dog sniff it. I naively asked him, "What's in the jar?"

"Burnt fingers," replied the trainer.

"Oh Christ," I said.

As the day wore on, Chief DiMarco called in Con Edison to find out where the gas meter was originally located. They told us it was on the rear basement wall. The fire fighters had originally noticed that the meter was beaten off the wall. We thought we should direct our digging in this vicinity.

I asked the trainer, "How will we know when this dopey dog detects a body?"

"Don't worry, you'll know."

With that another pile of dirt was scooped up and deposited on the ground. The dog looked as if he were electrocuted. He became as straight as a ramrod. Then he dug into the pile and brought up a tibia in his mouth. He started to run. I asked the trainer, "What will he do with it?"

"He'll eat it."

"Oh shit," I yelled, "get that mutt! He's eating the evidence!"

We managed to capture the dog and got the leg bone away from him. As I was taking the bone to our car the onlookers were asking, "Mister fire marshal we've never experienced anything like that, can we see?"

"No you can't."

More dirt was brought up and the mangy mutt again looked as if he had just been electrocuted. We were a lot quicker this time in getting the skull he found away from him. Most of the flesh remaining after the explosion had been eaten away by the rats. I put the wire of the evidence tag through the eye sockets. Then as I was carrying it to the evidence can in the car I flipped it up in the air, as

•

you would a softball. On seeing this about 75 of the hundred or so onlookers screamed and bolted in either direction down the block.

Shortly thereafter, the lower half of a man's body was retrieved. The pants were still on it. In one of the back pockets was Pedro's wallet. In the other was a candle.

The police drove by again, and on seeing the grisly sight said, "We'll take it from here."

We told them what they could do to themselves.

Eventually Pedro's partner was extradited from P.R. After Wes arrested him, he confessed that he and Pedro were promised money by the store owner to torch the place. They went into the basement and beat the meter off the wall. Then they went upstairs and started a small fire. Following this, they went across the street, sat on a stoop and, with beer in their hands, waited for the results of their handiwork, but nothing happened. The fire went out. Pedro decided to go back in. Rather than curse the darkness, he lit one of the candles he had. The place, however, was now filled with natural gas. The rest is history.

The Red Caps did a very good job. As a matter of fact, I think they did too good of a job, because the City cut back on the Red Cap concept. The City has been sticking it to fire marshals for years, since I've been here, because many of the people in government are not too fond of marshals. They can't control the fire marshals the way they control the police department. There's no two ways about it. It just seems to be the nature of the beast. The cop is more of a politician. A fire marshal comes from being a fireman, who looks at the crazy side of things, and he's out to make arrests of people that many police officers wouldn't. You can take the boy out of the jungle, but you can't take the jungle out of the boy.

I find being a marshal is very stressful. I don't think most guys ever get over the stress or the trauma of it, but you develop certain mechanisms to at least cope with it.

•

Still, we have a lot of guys who, after spending so many years in the firehouse, last only a year or so in the marshal's office before going back to the firehouse. Because as crazy as the firehouse is, there's still a much bigger support system there, and the black humor, which firemen are famous for, has its benefits. In the firehouse your lieutenant or your captain made all your decisions for you. As a marshal you're making decisions on your own. You're also dealing with the criminal element.

As a marshal you are trying to hurt that individual by incarcerating him and getting him convicted of the crime of arson. And you have to realize that's a dramatic change after coming from the rank of fire fighter where your whole act was to help: To help the stab victim, the rape victim, the fire victim. So I would say it's a period of adjustment of several years before you start to feel comfortable with the job—if you ever really feel comfortable with it.

When you get a case, you can smell blood, because you know this mutt has burned a building down and you pursue it like a hound dog. But you have to watch that you don't let your zeal run away with common sense. You can't let your zeal override this individual's basic rights under the Constitution.

When you go to the district attorney to present your case, you should be presenting all the evidence you have for this guy to be locked up. You also have to be open-minded enough to produce any evidence you have that indicates that this guy might have an alibi, because looking for the truth is the most important thing. But when I've got a guy who I am convinced is one of the happy match brigade boys, I'm going to go after him. I hate him and I'm going to do everything legally possible to nail him. I think this zeal is something that the fire fighters bring to the job, because practically every marshal has been to firemen's funerals. Every marshal has seen the burn victims, or has been burned himself. They know the pain.

There are two cases that come to mind. The first one

•

began when we got a call from the morgue. The morgue stated that they had an individual who had been burned five days ago, and they were making it into what we call "a medical examiner's DOA."

We go down to the morgue and we look at the body. His hair is gone. His eyelashes are gone. He has second- and third-degree burns, and he has inhalation burns. The M.E.'s report stated that Harry was burned in his backyard in the Country Club section of the Bronx.

We go to the address they have listed for him. The house is a beautiful brick home, no indications of fire. We walk inside. The bride is getting ready for her formal duties at the wake. She has her hair popped up and she's getting her mascara on. The father-in-law is scurrying back and forth, very nervous about us. We're here just to find out how old Harry got burned. The father-in-law says it happened right outside.

"Okay. Would you show us where and tell us how this fire burned poor Harry over 90 percent of his body."

"It was the lawn mower."

"It was the lawn mower?"

"He was filling the lawn mower and it exploded. The gas can exploded."

"Geez, that could certainly burn a guy over 90 percent of his body, if he's pouring a gallon of gasoline and the lawn mower did ignite it and blew the can up. He would definitely be covered with flammables. Well, where did that happen?"

"Right there."

He points to a spot that looks like the green on a golf course. The lawn is manicured. The grass is about an inch and a half high. It is beautiful.

"Where did this happen?"

"Right there," he said as he's pointing to the grass.

Now, if you've got a fire that burns a guy over 90 percent of his body, I would think that a couple of blades of grass might be a little singed. I'm now on my hands and knees

•

sniffing for flammables that might still be on the scene, but there's no odor and there's no observable indication of fire.

I said, "Well, I'll tell you what, you show me the lawn mower that he was allegedly filling."

"I threw it out. I was so upset I threw it out."

"Oh, that's understandable."

"The department of sanitation picked it up, yesterday."

I said, "They don't pick up on that day."

"No, you're right, I threw it in a lot."

"Well, okay, that's understandable. You were pretty upset. Your son-in-law was burned to death and you threw it in the lot."

"Where's the lot?" I really wanted to get a look at this alleged lawn mower.

"I don't remember."

"Well, this Country Club section of the Bronx is very nice, but there's one lot down the block. I know about three blocks down there's another lot, but they're building on that one. Did you throw it in that lot?"

"No, it wasn't that lot."

I asked, "Was it in the area?"

"Yeah, it was in the area."

"Well, if it's in the area, what lot was it?"

It was obvious to Richie and me that we're getting the bullshit story. We had to write up a report stating that we found no indication of a fire at this particular address. We couldn't do a "cause and origin" examination, because there was nothing to examine. We had to leave it open. There was just no indication of fire, but we have a guy who's burned over 90 percent of his body, and his whole clan is getting together and they're ready to have a big funeral with the big clock stopped at such and such a time when good old Harry died. I'm feeling a little lost and frustrated. We didn't know what the hell to do. We're at a dead end. We go on to the next case.

Not three weeks later we get a call from the Rockland

•

County, New York fire investigators. They want to come down and talk to us about a certain individual. We meet with them.

Apparently, what had happened was that Harry was a paid torch, but he wasn't too professional. He went up to a pizza parlor in Rockland where he was hired by the owner to burn the place to the ground. Harry, being a paid—but not a smart—torch, marches in there with 10 gallons of nice, high-octane gasoline. Harry commenced to pour the gasoline all over the first floor of the pizza parlor.

Why he's not smart is because a professional torch would use kerosene, never gasoline. Why is that? Because kerosene is a lot less dangerous to the guy who's using it as a fire starter. It is less volatile and has a higher flash point. It takes a while to get kerosene going, but when kerosene does get going it's really tenacious. It really bites. It gives you a hotter fire. Gasoline is a very, very dangerous flammable to use.

What Harry the nonprofessional doesn't realize, as he's pouring the gasoline all over the first floor, is that gasoline vapors are heavier than air; therefore the vapors from gasoline, which are actually the things that burn, are working their way into the stairway and into the cracks of the building and down into the basement. Lo and behold, what's in the basement waiting for those vapors to come down?— the gas burner, which has an open pilot light. Well, when the vapors finally get down there, the place goes up with Harry still holding the gasoline can in his hand. Harry had poured so much gas that the explosive mixture became so heavy that the fire burned, but then just about extinguished itself because it burned up all the oxygen. There wasn't a tremendous amount of damage.

The Rockland investigators had a picture of the door to the street. It was one of those glass doors with a metal frame around it. On the door you can see the key sticking out that the owner had given Harry to get in. When Harry

•

got in, he put the key into the backside of the lock and locked himself in, so that nobody else would come in. You could see that the whole door was covered with soot, and Harry's hand prints were all over it. While he was on fire he's fighting to get the key. He finally finds the key, unlocks the door, and out he goes.

The son of a bitch with 90 percent burns over his body manages to make it from Rockland County to Jacobi Hospital in the Bronx. That's got to be at least 30 miles. However, 5 days later he died.

Here's a case where you don't know what you have, because all of a sudden he came in from Rockland. But by that time any leads are cold. You can't find out anything. We didn't have the means to get to talk to that guy in the hospital. That could have turned the case and he would have given us the guy who paid him, because the fire took hold of Harry. We would have said to him, "Harry, you're going to die. You are going to die right there in that bed. Now don't you think you want to make peace with God and give us the guy who paid you?"

The other case was a fire we had on Nagle Avenue in northern Manhattan which turned into a third alarm. It occurred in an old bakery in a row of "taxpayers." The guy who made the fire went in and spilled 10 or 15 gallons of gasoline in the place and then he lit a match, turned and ran out the front door, which was open for him. When he got out the front door, there was a cop car coming up the block. So he turned around and ran back through the building to escape out the rear. But he ran back through his own handiwork. The witnesses that we interviewed heard an ungodly scream in the backyard the night of the fire. It was this guy running through the backyard on fire.

We thought for sure this was going to be very simple, because if he got burnt maybe he had a cohort. Maybe the cohort is still in there. We spent three days digging that place out to make sure there were no bodies in there. The cops had seen him running back in, and although the

•

witnesses heard him in the backyard, nobody saw him run out. There were no bodies, so we went to all the surrounding hospitals. We still felt that this is going to be a piece of cake. He has to be burned over 100 percent of his body.

The hospitals, however, stonewalled us. They would not let us look at their records. They would not let us look in their burn units. They wouldn't give us any information whatsoever. That case ended up getting closed as an incendiary fire, but we were stymied in the investigation because we were stymied by the hospital.

Those two cases led me into persuading John Regan, the chief fire marshal, to let me work on developing legislation which says that in the event somebody comes into a hospital with the burns I described, the New York State Office of Fire Prevention and Control would be notified, who in turn would notify the respective agency that was in charge of investigating fires in that particular town or county.

The Burn Reporting Law parallels the gunshot law, which says that if an individual comes in with a gunshot wound, the doctor is mandated to report it to the police. Now we have a law on the books that if anybody comes in to the hospital, clinic, or doctor's office with second- or third-degree burns over more than 5 percent of their bodies, pulmonary edema or laryngeal edema, the doctor is mandated to report it.

The law was signed four years after those fires. It took four years of fighting to get the legislation passed. We finally were able, with the help of Barbara Shulman, a fire and arson project coordinator of a community-based organization, to get the Assembly and the Senate to pass it unanimously. We finally got it signed by Governor Cuomo in June 1985. It, however, didn't take effect until November 1, 1985.

On November 3 we made our first arrest as a result of it. Louie Garcia, a fire marshal, made the arrest of a young lady who was very upset. She came home and found her

•

man asleep in bed with this strumpet from next door. So she methodically took one of these big three-gallon stew pots, filled it with water right to the top and got it boiling. Then she put on the big oven gloves and picked up the pot of boiling water. She went in and poured it all over her boyfriend, while he was lying in bed. She poured it on his face and on his balls just to show him that she was a little upset. He died of scalding wounds.

I personally feel that we're only seeing about half the burns that fall within the parameters of the bill, but we are working very closely with the medical field. Because if a doctor does not report the burn, he can be charged with a Class A misdemeanor, which could mean he'd spend up to a year in jail or pay a $1000 fine. I know when we lock one of these nonreporting doctors the word will travel very fast in their circles and we'll get 100 percent compliance.

There have been a number of other very good arrests since this legislation came into being.

On the morning of February 24, 1986 a full first alarm assignment was sent to 204 Eighth Avenue, Box 595, in Manhattan, for a fire in a five story, brick, old-law tenement. The fire was in the basement and up the interior stairs. A second alarm assignment was needed 15 minutes later to help put it out.

When Fire Marshals Carmine Palopoli, Charlie Gaglian, Tommy Buda and Supervising Fire Marshal Cecil Maloney got there, they found some very angry tenants. The tenants told them that there was a rent strike going on and the landlord wanted them out of the building.

Inside the building, the marshal detected an odor of gasoline. In the front room of the cellar there was no ash residue to account for the heavy volume of fire, and at the top of the interior entrance to the cellar a door and its frame were blown off the wall.

A short while later the marshals got a report of a man being admitted to a hospital for burns. He matched the

•

description of someone seen running from the building just after the fire started. They paid Dwayne a visit. He denied knowing anything about the fire, but a check of his criminal record showed that he was wanted on an outstanding warrant. He was arrested and eventually transferred to the infirmary at Riker's Island.

Dwayne eventually dropped a dime on his associates because he was convinced that one of them "ratted him out." It seems that the landlord, Gregory Gelman, a 39-year-old Russian émigré, and his partner paid Dwayne and three of his friends to go in and torch the building.

After collecting a great deal of evidence and conducting many interviews, an arrest warrant was issued for Gelman. When they went to his apartment, he wasn't home.

They then went to his girlfriend's apartment in Greenwich Village, but they didn't have a warrant to enter her place, although they did have positive identification that he was in there. Four marshals staked the place out throughout a rain soaked night. At dawn they called the apartment and said that Gelman must surrender. This was a bluff.

After consulting with his attorney, a former Bronx A.D.A., a red-eyed Gelman came out.

On October 6, 1987, Justice Henry Atlas of the New York Supreme Court found Gregory Gelman and his partner guilty. The other four were also found guilty in subsequent trials.

Gelman was also a concert violinist—he gave a Carnegie Hall recital in 1982 that critics described as exhibiting "an accident-prone technique." His next accident will probably occur in jail.

Gelman was convicted of first-degree arson, second-degree conspiracy, and first-degree reckless endangerment, and given a 15-years to life prison sentence. His partner got 1⅓ to 5 years for second-degree conspiracy. Dwayne is doing 3 to 6 years for third-degree arson.

•

After my success with the burn reporting law I decided to deal with the specter of Pepito's death that had been haunting me for almost 10 years. In July of 1985, I approached Chief Fire Marshal John Regan with an idea for a juvenile fire setter intervention unit. The first thing the chief said to me was, "I don't want to be locking up kids."

I knew how he felt, but I also knew that we needed an alternative to incarceration. After I explained my position to him, the chief gave me the OK to get a program off the ground, and after four months assigned Charley Wagner to help me. Charley had only been a marshal about a year and a half, but he had an excellent reputation as a fire fighter and had worked in a number of busy companies. I put a lot of faith in guys with that kind of background.

After much work, and a lot of help from Gerald Bills and Bob Crandal, who run the Rochester program, we were given the go-ahead for a one-year pilot program in the Bronx.

We modeled our program somewhat after the very successful one in Rochester. It is a sophisticated program that requires the cooperation of many other agencies, but it is able to handle not just the curiosity fire setter, but those kids that might have other serious problems that cause them to start fires as a cry for help.

Several months before the program was set to start, on the morning of October 11, 1985, Ladder Company 42 responded to Box 2261 on Prospect Avenue in the Bronx. When they arrived, there was smoke pushing out of several windows of a fourth-floor apartment in an occupied tenement.

Captain James F. McDonnell led his forcible-entry team, fire fighters Peter Bielfeld and Daniel Saitta, to the fire floor. Searching for trapped victims, they crawled 40 feet into the burning apartment—down the interior hallway, through the living room, and into the master bedroom. But the situation was terrible. The master bedroom was filling with superheated smoke. A flashover was imminent.

McDonnell turned and shouted through his mask, "Get out!"

As the room erupted into a fireball, McDonnell shoved Bielfeld to safety. But the fire caught McDonnell.

The nozzle man opened up the line and knocked the fire down. The other fire fighters managed to get back into the bedroom where they found McDonnell lying in the rubble. His mask was melted on his face.

As he was being rushed to the burn center at New York Hospital he suffered a heart attack. They gave him CPR. He kept coming and going. Nine days later James McDonnell was dead.

The fire that killed Captain McDonnell was started by a four-year-old boy. It was the kid's third fire that required a fire department response.

We named the training component of the juvenile program in honor of Captain James F. McDonnell.

# 2

## SUPERVISING FIREMARSHAL

## WILLIAM MANAHAN

'll tell you how naive I was when I went into the BFI, in 1979. After the first couple of months we went down to alphabet land in lower Manhattan. I can't remember the street, but the building was across from where the Hell's Angels motorcycle gang hangs out. We didn't think anything of it.

We went into this six-story elevated building. The elevator was broken so we walked up the stairs, and the first thing we ran across was a guy shooting up heroin. We're in civilian clothes so he didn't know who we were. As we climbed past him we saw that he had the rest of his works—a bottle cap, cotton, a small bottle of water, a glassine envelope and a small lit candle—spread out on a step. We snickered about it as we went up to the floor where we were to investigate this vacant-apartment fire that happened the day before.

Since we are required to get an interview as close to the fire as we can, when we got up to the fifth floor we knocked on the door of the apartment

•

across the hall from where the fire occurred. We knocked a couple of times before this guy answered.

He was white, with long hair, emaciated, tattooed. It looked as if he was a serious substance abuser. He walks out into the hall, but he doesn't close the door behind him. He leaves it a little ajar. He asks us, "What's up?"

As soon as he said that we hear a shotgun being racked inside his apartment. We figure that we're in over our heads. But I'm talking to him about a fire that happened the day before in the vacant apartment across the hall, which he claimed he didn't know took place and didn't care that it did.

He was trying to figure out if we were cops or if we were trying to rip him off. What he had in the apartment of course was drugs. Maybe the place was some sort of factory.

The biggest thing my partner and I knew was that we were in some place we shouldn't have been. Though we have radio contact with the base through our Handie-Talkies, there is only a light-duty fireman on watch, and if we call in for help all he'll do is telephone the police operator at 911—as long as he has not gone to the kitchen for a cup of coffee. We're really out on a limb. We're by ourselves, so we figure the best thing to do, since we did the physical, is just disengage.

I talked fast. I said, "Look, we're marshals. We're here about the fire. All I need is a name"—I don't know why I went this far—"so I can type up a report." He gave me a Spanish name. It was obvious he wasn't Spanish. I couldn't remember the address of the building we're in, I was so nervous, so I asked him his address. He gave me an address in Germany. I took it, thanked him, and we walked out.

We went down to the 9th precinct. The desk sergeant sent me up to the squad and I told them about what went on in that building across the street from the motorcycle

•

gang. They said we have SOP orders that even uniformed cops have to go in doubled up. In other words, if a squad car goes to that building, he automatically gets a backup, so there's got to be a minimum of four cops that go in. The cop also has instant help because he's tuned in with his radio right to the police dispatchers.

On my very first job as a marshal I was out with a senior guy by the name of Tommy Buda. It was bitter cold. It was February. We went to a bar on Atlantic and Troy Avenue. It was "a threat to burn." They had ejected a patron out of the bar, but before he left he had threatened to burn the place down.

I remember Tommy telling me, "You just observe me and you'll catch on. This will be your training period." He started talking to these people. He invited them into the car one at a time. We had the heater on. Tommy is very methodical and precise. But I didn't understand the idea of becoming friendly by using small talk with witnesses, so I became overly anxious. I was like a horse in a paddock ready to break away. I kept on injecting questions from the backseat, where I was sitting. Tommy was in the driver's seat, and the witness was in the passenger seat. At one point Tommy asked the passenger if he would mind stepping out of the car, and I thought we were going to have an insightful conference. Maybe one of my questions picked up something. After the guy left the car, Tommy turned to me and said, "You open your fuckin' mouth one more fuckin' time and I'm going to jam my fuckin' fists down your throat."

I said, "What? What?" I was taken aback.

He said, "I can't concentrate. Don't ever ask questions when I'm talking. When I finish, I'll ask you if you have questions and I'll give you as much time as you want."

The guy came back in the car and Tommy finished talk-

•

ing to him. Then he turned to me and said, "Bill, do you have any questions you want to ask this guy?" I refused to talk. I was upset.

But that was Tommy's way of doing things. Before I went out with Tommy I was told how good he was; but I was indignant.

After the tour I went for coffee with the other new guys. I told them what Tommy said. Then I realized it was worth getting yelled at to learn something important. During my 10 years in the marshal's office I found that one of the hardest things to do is to calm a new guy down so he'll not break your train of thought. I also found out that everybody gets yelled at for one thing or another.

Another fault of new people is that they put too much energy into a job that's not going anyplace. There is a cost/return equation in every investigation, and as you mature in the job you learn which ones to stay on and which to cut loose. An example of the latter was our threat to burn case. There was no follow-up on it because we had no name for the person who made the threat, and only a vague description. The neighborhood had many transient people, and we had about four other jobs to do that night.

At that time we covered half the city. In other words, the fire marshal assigned to the Brooklyn base could respond to midtown Manhattan—from Fifty-ninth Street down, all of Brooklyn, all of Staten Island and the Rockaways out in Queens. You could spend a lot of time on the road, so you had to be very, very conservative with how much time you spent on each job. Often a new guy would spend too much time on one job, so he'd wind up bunching up at the end of the night. You had to learn to pace yourself.

One night I was assigned to go out with two senior guys, John Knox and Jimmy Callender. I had hardly any experience, but we had an investigation on Eastern Parkway in a building that had a fire in an apartment. We had to interview the tenants of the building. When we started, John Knox said, "Billy, this is yours."

The first door I knocked on, an old black man opens. Knox tells me, "Show him your badge." I pull out my badge, and all the change comes out of my pocket and falls all over the floor of this elderly man's apartment. As I'm fumbling with my badge, he closes the door in my face with all my change inside. Knox said, "Come on, Billy, let's try this again." Needless to say, I was embarrassed. To this day, he reminds me about it.

When you arrive at a job, the closer you come to the actual fire the more it's a mess, especially if it's a job that's going to go someplace, like a homicide, or where you have multiple 1045s. That's where people are burned, or where you have what they call a newsworthy fire, something that the newspapers are very interested in, or a fire in a religious institution, where you have to do a lot of extensive following up.

In order to grab all the information you need, you have to do a balancing act between all the people on the scene before they are gone. You have a myriad of emergency personnel there, including the police department, the emergency medical service, and the Red Cross who get involved with the civilians who need relocation, or are in a state of shock because they lost all their possessions. You may have insurance adjusters there. Maybe you might even have a suspect, who people are accusing of making the fire, that you want to focus in on. In between all that you have to do a physical to prove that this fire is either accidental or incendiary. It's chaotic, to say the least.

You also have to catch as much information as you can from the fire fighting force that just went through this horrendous operation. Well, I'm not an ex-fire fighter. I *am* a fire fighter. I know what they went through, especially under severe climate conditions, like excessive heat or cold. These guys are beat. You can see they're expend-

•

ing their last physical effort just taking up the hose or putting the tools and the ladders back on the truck.

You come in all fresh and bouncy and bombard them with a hundred questions, but you have to develop an approach that doesn't piss people off. You have to identify with them. You might say, "Look, I've been where you are and I know it's hard, but my job here is a backup for you guys so this doesn't happen again. If I can just ask you to extend yourself a little bit further. . . ." Usually it works, because firemen are good people.

I want to catch the fire fighters while the incident is fresh, because if a case goes to court their testimony is very critical. They can tell you if the doors or windows were opened or closed. They can tell you the condition of the fire when they arrived. All of which is important to your physical determination.

One of the most critical things we have to do is establish a linear sequence of events, and the way we do this is to begin when the box is reported to the dispatcher's office. The fire department also documents, by radio signals that are recorded in the dispatcher's office, when the companies first arrive. So we know the exact amount of time from the moment the fire was reported, to the time the firemen arrive. Though we have a three-minute time gap, the observations of firemen can still tell us the progression of the fire.

The significance of locked or unlocked doors comes into play when we take testimony from a suspect involved in a commercial fire where the guy is trying to pass it off as "somebody broke into my store and must have robbed it. Look, my cash register and my money box are gone. All the merchandise is gone, and now there is a fire."

Well, if a fireman had to come in and pop about four big locks off a roll down gate in the front, and the back door is bolted, and the skylight is intact, and the fire wasn't visible from the outside until the firemen opened up, then you can safely say in court, "If that was the case, how did

•

the burglar get in?" That indicates that the fire may have been set prior to this door being locked, and so it might open up the possibility to prove exclusive opportunity.

If you know firemen are opening up windows and doors in that initial period before they can extinguish the fire, especially in difficult fires, then you can correctly explain some fire patterns that you might otherwise misread. For example, if you see fire traveling in an unusual way, it might be due to the fact that sometimes people, to effect a lot of damage, will leave interior doors open or windows open, or put holes in the roof. So, it's very important that you get information from the fire fighters about what they did.

It is a toss of the coin as to where you should put your emphasis when you first arrive. Should you put your emphasis on the physical, because you don't want things to change in that area? Or do you put your emphasis on getting statements from people, because people disappear, or they change their minds? Or worse than that, they get talking to each other, and start influencing each other's stories until the stories start melding into one. Consequently some critical piece of information could be left out. Or a person might be embarrassed to say something, like he smelled an odor of gasoline, when the other people say no, we didn't smell it. He'd be reluctant to tell you that because he wants to be with the consensus of what the crowd is telling you. And then of course you've got the people who are responsible for the fire. You want to grab them before they can reorganize their thinking, so you can lock them into something that you might hang them on later.

The people who give us the most trouble are the ones who probably suffer the most during the fire. It's hard to get them to sit down and talk when everything they have is gone and they have to worry about where they're going to sleep, or that they have no insurance. Many people come from poor families where they can't impose on their

•

relatives. They're wiped out. They're like war victims, and you're trying to get information that seems incidental. "Why is this guy bothering with this stupid shit when I lost everything?"

On the extreme end is when somebody has just lost a member of their family, or worse, their child, and you have to talk to them. It's not that they purposely give you a hard time, but we are on two different missions. They're in mourning and shock, and are part of the panic and fire. You are not. In fact, often there is a little resentment, because these people might be wet, they might be dirty, they might be cold, and here you are, dressed appropriately, fresh faced, and asking questions about what color was the smoke, what did you do, how did you know about the fire. It's hard for them to forget their present predicament and recall those things. You're asking them not only to participate, but to drop their feelings and go back mentally to a very traumatic situation, and it's very, very hard for them to do that.

Of course, we want to know what color the smoke is because that would give us an idea if a chemical was used to start the fire, or if there was a particular substance in the fire load of the building. Sometimes, a gas burns with a distinctive color; for example, carbon monoxide building up in an enclosed area would burn with a blue flame, or sometimes a little flame with heavy black smoke is an indication of a hydrocarbon being used, in other words, a flammable liquid. This information can help tell the progression of the fire. If the guy says, "There was no flame, but a lot of smoke" you can say, "Well, maybe that was a phase of the fire where it was smoldering."

In fact, in one of our cases the progression of the fire was critical to its outcome. On the morning of March 27, 1987, my partner Donald Green and I were called to 390 Gates Avenue, Gates and Nostrand, in Brooklyn. The fire department had an "all hands" fire. They called us because they found a dead 29-year-old, Hispanic woman lying faceup in a burning bed, and they couldn't determine the cause of the fire.

The victim's name was Elizabeth. Her sister Maria and her brother-in-law lived in the same building. They told us that the routine in the morning was for their 9- and 10-year-old daughters to come down from the ninth floor and knock on Elizabeth's door, and she would take them to school. At 7:15 this morning, the two nieces came down and knocked on her door, but nobody answered, so they went back upstairs and told their mother. The mother said, "You go back down and knock on your aunt's apartment. Ring the bell and knock hard, because she's there." They came back down, at 7:25, and they rang the bell and knocked really hard on the door, but still nobody answered. Then they went down to the security man in the lobby, and he buzzed and buzzed the apartment on the intercom, but still no answer. The kids went back up to their apartment while their father got ready to take them to school.

A short time later on the second floor, the smoke alarms in several of the apartments started going off. One tenant, a guy named Vasquez, comes out in the hall to see what is going on and finds it filling with smoke. He immediately starts knocking on the other tenants' doors. When he knocked on apartment 2A, no one answered, but puffs of smoke came out of the cracks around the door. Now he pulls the interior fire alarm next to the elevator, which also alerts the fire department.

We arrived at the scene and went up to a second floor apartment. As soon as I looked at the fire room, I knew there was something very wrong. It was a very strange fire. There was a massive disruption of the material in the mattress and the box spring, but very little fire damage in the room itself.

The fire fighters had already put Elizabeth's corpse in a

•

body bag. I requested that they bring the body back into the apartment from the public hallway, where it was waiting to be taken to the morgue.

We then unwrapped it and examined it in detail. I discovered that this victim had her shoes and her jewelry on. There were also remnants of her clothes and her face was covered with some sort of cloth. It was evident that she was very passive to the fire. By passive I mean that she was in the supine position on her back. Normally the body recoils from a fire. If you had a fire in your bed, unless you're really unconscious, you would recoil from the fire, and even if you woke up for only an instant, you would try to get away from the fire, or you would try to protect your face. Her arms were raised off the bed, by the contraction of the muscles due to the heat of the fire. Her nostrils were closed up, but her mouth appeared clean. We were unable to examine her throat.

Normally when you sleep in a double bed you will sleep on one side of it or another. It was evident that she had been lying in the center of the bed. Subsequently I studied how to get someone in that position. I have a 15-year-old son and I practiced putting him in bed. When you put a person that weighs at least 125 pounds in bed you put your knee on the edge of the bed and you set them down in the middle of it.

We concluded that before the fire she was either dead or unconscious, and she was laid out with her arms down at her side.

We were also able to prove, though it was extremely hard to do so, that the fire was set at multiple points on the mattress. We did this by analyzing the remains of the springs. In a normal fire the body of the victim acts like a fire stop. For example, if the fire started on the left side of the victim you would have an uneven progression with more severe annealing of the springs on the left than on the right side of the victim. But what you had here was complete material destruction of this mattress evenly

•

around the body. In fact, the victim was evenly burned on both sides of her body, and there were also signs that a big cushion from a bamboo and rattan "papa-san" chair was over her face, and that too had been on fire. Then of course the sharp lines of demarcation between burnt and unburnt areas surrounding the bed told us that it was initially a rapid-burning fire.

Since we thought this case involved a possible homicide, three detectives, Lettau, Cavute, and Gibbs, from the 79th Precinct were also assigned to the investigation.

The principal suspect was a guy named Smith. He was her common-law husband. Elizabeth's relatives were really fine people, and they were very upset that she had taken up with this guy. We learned that he beat her up on several occasions, and on New Year's Eve he had blackened both her eyes. The security man told us Smith left for work at 7:50 that morning. Elizabeth's relatives informed us where he worked.

Mr. Smith was picked up at his factory job later that morning and brought to the 79th Precinct for questioning. He volunteered to talk to me, but he couldn't make up his mind about certain facts. That became very important in my testimony in court. He told me that there was a digital clock right by him in bed. He was late for work. He woke up, for some unknown reason, at 7:45, put on his pants and shirt, walked to the bathroom adjacent to the bedroom, threw cold water on his face, stopped in the hallway of the apartment, and talked to Elizabeth. At first he said that she was sitting up in bed smoking crack when he left.

I said to him, "You're telling me that you spoke to her from the hallway? That means that the bedroom door must have been open." I knew the bedroom door was closed from the fire burn patterns.

He paused for a good 20 seconds then looked at me and realized that something was wrong. He said, "I want to start over."

I said, "Sure."

•

He started over. He said he woke up at 7:45 and went to the bathroom. He closed the bathroom door behind him, because he wanted to turn the light on in the bathroom, but he didn't want to wake Elizabeth.

I said, "Oh, she was sleeping when you left?"

He said, "I want to start over again." Then he said, "I jumped up and put on my clothes. I didn't see Elizabeth. I don't know anything about the door. I just left."

He said he left the apartment at 7:50. The fire alarm was received in the Brooklyn dispatchers' office at 8:08. So you've got an 18-minute gap for this fire to consume the mattress, leave an enclosed bedroom, travel about 26 feet down to the front door, and for the smoke to get into the public hallway and travel another 40 feet, until it set off smoke alarms at the other end of the corridor. Plus, the firehouse is right around the corner and the alarm came in at that time of day when everybody was on the apparatus floor. They got the signal at 8:09 and were there within 2 minutes.

What was critical about talking to the firemen was that the fire fighters coming in with the hose line could not find the fire, because there was no visible fire. Fireman Michael Banker from 102 Truck put up a portable ladder before the bedroom was vented and before the engine company got up to the fire. He was able to look in the window and see the body on the bed engulfed in a smoldering fire. We estimated he looked in at about 8:14.

The other thing is that a lot of the clothing was strewn around the floor; it was even strewn behind the bedroom door. That made the room relatively airtight. The fire burned out the oxygen in the bedroom. However, simultaneously, it was burning through the top of the double plywood bedroom door. So as the fire was burning itself out it created unequal pressure between the bedroom and the hallway, and oxygen could not come back in to feed the fire, because there was too much pressure from the smoky gases inside the bedroom. The fireman from Ladder

•

102 told me that once he took out the window the fire lit up. Then we knew for certain that we had an oxygen-starved fire that burnt itself out, yet took enough time to waste away all the material in the mattress. Therefore, it had to have been going far longer than the 18 minutes that would be consistent with Mr. Smith's story.

We conjectured that Mr. Smith used a pint-sized butane torch to set the mattress on fire, because, in the absence of flammable liquids, a good heat source was needed to get the fire going the way it did. The victim's relatives also said that he always used the torch to smoke crack. On the day of the fire it was conspicuously absent.

When we had Mr. Smith at the precinct, it wasn't until about 10:00 at night that we were able to run his criminal record. We found out he did 7 years of a 10-year sentence at Attica, in upstate New York, for strangling his first wife in 1965. After he strangled her, he put her in the trunk of a car and set it on fire.

The ADA on the case became very interested in Mr. Smith's history. We started turning up indications that he had trouble with ladies, especially the ladies he was living with. He was the type of guy who had a pimp personality. He was able to pay a lot of attention to his women. He slept little, and was into drugs. However, for some reason, every once in a while, this nice-guy routine would break, and he would become very violent.

We found out that he left a town in North Carolina in 1960, where he was the chief suspect in the shooting death of his 15-year-old girlfriend. He came up to New York, where he did seven years for killing his first wife. He then moved to Newark, New Jersey. He was also the suspect in the death of a Newark lady.

Plus, some woman from Newark later told us a story that she was also his common-law wife. She had literally jumped through the glass of a second-story window to get away from this guy. And the reason she had to jump through the window was not to save her life, but she knew

•

if he continued to strangle her he'd have to kill her daughter, who was also in the apartment. She broke both her legs. He was charged and arrested, but somehow he never did time for that.

He then came back to New York and lived with a few ladies that we were able to trace. They also reported that he had a violent temper. He broke up with one woman and right after that her car was set on fire.

At one point I said to him, "By everybody's description Elizabeth was very pretty. She kept the apartment real clean and did everything for you. This must have been some loss for you. But since we picked you up at work, you never asked much about her, nor do you look particularly upset."

He smugly replied, "That's the way it goes. You have to move on."

We wanted that guy so badly we could taste it.

At 4:00 A.M. on March 28, after trying to bluff his way through a videotaped interview with assistant district attorney Ann Rost, Mr. Smith was arrested on a charge of second-degree homicide.

Three days later, we went before the grand jury and got an indictment. Mr. Smith pleaded not guilty. However, he could not make bail, so he spent the next year in jail waiting for his trial to start.

The medical examiner agreed that the victim was dead prior to the fire, because her larynx and lungs were clean, and there wasn't a high level of carbon monoxide in her blood. However, because there was cocaine in the victim's body, she couldn't rule out a drug reaction, even though there was no indication in the autopsy that Elizabeth's heart fibrillated. The cause of death was given as either drug reaction or suffocation. I believe she was suffocated with the pillow.

Because the medical examiner could not prove the actual cause of death, we were unable to link Mr. Smith with the

•

murder of Elizabeth. However, we were able to go to trial on a charge of second-degree arson.

At the start of the trial, before the jury was brought in, they conducted a Huntley hearing, because Mr. Smith's attorney filed motions to have the statements he made to us thrown out. Since his statements were critical to the case, they attacked me. They tried to infer that we threatened Mr. Smith and therefore coerced him into saying the things he said. The judge ruled in our favor that Mr. Smith's statements were admissible.

Right after the ruling, the jury was brought in and the trial started. The trial lasted three days, but there was a problem. Because Mr. Smith was only being tried for the crime of arson, no mention could be made of Elizabeth's body. The ADA had tried to introduce our photographs from the forensic unit, but Judge Thadeus Owens ruled that they were too prejudicial, because they showed how badly Elizabeth was burned. The ADA then took a razor blade and cut out Elizabeth's image, but the judge still said no. Thadeus Owens is a black judge who worked his way up. I found him to be tough, but he was also very fair.

I testified late one afternoon. I'd rather be fighting fires than testifying in court. That walk from the back of the courtroom to the witness stand I call "the last mile." It is just as anxiety-provoking as going to a fire. In the court it can get pretty abusive. The defense attorneys call you a liar. They attack your credibility. They look for fault in everything you have said or done on the case.

Before I went on the stand, we first put up there all the fire fighters that I had interviewed at the scene. Juries are usually impressed to see a uniformed fire fighter get up there with his display of medals, plus fire fighters historically have a good rapport with the public. Since we're in civilian clothes, we sort of ride on their coattails.

Because fire marshals do a physical examination, when I was describing the points of origin of the fire I was able

•

to make mention of Elizabeth's body. Well, now we had a jury box full of people with confused looks on their faces, because this was the first time they heard about a body. However, early the next morning, we were able to prove to the jury that Smith was in the apartment when the fire started.

When I finished, the medical examiner went on the stand and testified. Then the judge charged the jury and released them for lunch. Within 45 minutes after returning from lunch they brought in a guilty verdict.

Mr. Smith is now in an upstate prison doing 8⅓ to 25 years for arson. However, because of the time gap between the death of his first wife and this latest conviction, Mr. Smith is not considered to be a predicate felon, which would require him to serve the full 25 years. Now, with good behavior, he can be out on parole in mid-1995, and the first thing he will do is find himself another lady.

All in all though, it was a good team effort involving the marshals, the police department, and the Brooklyn district attorney's office.

In fact, there's a rule of thumb that you get treated by other agencies better than you do your own.

I've never, never had an incident where I had trouble with the NYPD. In fact, I think I get better treatment than they give their own.

After you make an arrest you have to go to central booking. It is a maze. Your first time down there you don't know what the hell you're doing.

I made one of my first arrests in Manhattan. I arrested a 19-year-old private security guard for setting a series of fires in an exclusive hotel, still under construction, but partially open. It was the Helmsley Hotel on Forty-second Street.

By the time we got him videotaped, making his statements to the ADA, it was 1:00 in the morning. Then I had to take him to Manhattan central booking.

•

When I got there, I didn't realize it, but I was responsible for fingerprinting my own prisoner. I was given the cards by the desk officer, and he said, "Fingerprint him and bring the cards back to me."

That was the first time I ever fingerprinted anybody; most of the time in Brooklyn they do the fingerprinting for you.

I brought the cards back, but they were all smudged. The lieutenant ripped them up and gave me a new set of cards and said, not in a fatherly manner, "Here, do the fucking things over again."

I came back to him, but they were messed up again. He handed me the third set and said, "You fuck these up and I'm throwing you and your prisoner out of here."

He put the pressure on me, and I screwed them up again. I walked back to him and said, "Lieutenant, I'm sorry. I need help."

I had forgotten to display my badge. He looked at me and said, "What command are you in?" He was ready to take my head off.

I said, "I'm a fire marshal."

"Oh, a fire marshal, don't worry." Then he rousted some guy from eating his meal and said, "Do the fingerprinting for the fire marshal."

It wasn't 20 minutes later that some young cop came in and screwed up a set of fingerprints. He got his head chewed off, and they didn't roust anybody to help him. That's the way it is. I'm an outsider. I guess I'm expected to make mistakes.

Each borough has its own central booking and consequently each central booking has its own personality. They have the same rules, regulations, and requirements; but they all seem to be different.

Brooklyn is very, very busy. It's like a crowded supermarket, but the guys who run it handle the situation well and get everything done as quickly as possible.

•

Manhattan is very colorful: A vast variety of people. I've seen everything there from female impersonators to midgets.

One day, when I made an arrest for another hotel fire, I went down to central booking in Manhattan.

Central booking processes the prisoners and then farms them out to different holding areas within the city to await their court appearance on the next day. Something went wrong with the farming-out process and there was a back-log.

They have this big holding area, about 20 by 30 feet, and in it are some benches which are bolted down.

Right before they farm the prisoners out, they handcuff them to a 12-foot chain. Eight or 10 people are on this chain, along with this guy who looked like Mr. T. A very agitated, big, black guy, rippling with muscles. He was the first guy on the chain and he was pacing up and down the 30-foot length of this cell, almost at a jogging pace, and everybody that was chained to him had to follow and be quiet. They were afraid to say anything to this guy.

I was there a couple of hours, and he just kept on pacing. Finally he sits down on one of the benches, and everybody behind him collapses wherever they could.

The cop at central booking, without missing a beat, said, "Hey you fag, couldn't take it? Got tired?"

So this black guy jumps up, all aggravated. The rest of them couldn't get on their feet fast enough. He starts to drag these people up and down the cell again.

Then of course there's the misery part of it. We often think in terms of black and white, that the people who are arrested are these really bad people, but sometimes the people fall into the realm of being victims themselves. I don't have any political agenda to justify this, but they're the people that for some minor reason or another wind up in central booking. They are the first-time people, like the suburbanite who got into a fight.

You can see how the process really affects them. In fact,

•

if it wasn't for the strict, conveyor belt routine, I think many of them would be overwhelmed by being in jail.

I always wondered what they experience at the end of the process, when what we call "logging in the prisoner" happens. That's where you turn them over to the cop, and he takes their shoelaces and their belt, puts them into a cell that smells like a urinal, and locks the door.

Sometimes it's really upsetting for me to look at their faces. Because as much as you're in an antagonistic situation with them, they don't want you to leave. Maybe I identify with them. I realize that some unfortunate situation could happen to anybody.

But I have to remind myself that I was on the line and I went into the fire, and then there are all the firemen's funerals I went to.


I had arrested someone for a fire at the Helmsley Hotel. At that fire, because of the heat, a 55-gallon drum full of number 6 fuel oil swelled up half again its original size, and was ready to burst just as the guys from my old company, 54 Engine, were taking the line in.

Every arsonist you put away protects a building, protects civilians, protects fire fighters.

I have never arrested anybody that made bail. They all stayed in jail until their cases were disposed of. A couple of people could have made bail, but on the advice of their lawyers they didn't. These were the people who had an emotional problem with setting fires in the hotels. In one case there was a subrogation suit going on.

What happened was that the Helmsley Hotel on Forty-second Street was under construction.

There are these hotel laws that stipulate that the management is allowed to open up one-third of the hotel and operate it for 90 days. Then it is evaluated, and if the operation is approved they can open up the rest of the hotel.

•

They have the initial grand opening. The mayor and the governor are there, and they have a second-alarm fire.

They also had a series of fires that preceded the opening and another series of fires afterwards. We thought they were having labor problems.

I was working with John Lagoff, and we were able to pinpoint a suspect who confessed to the second-alarm and a couple of the small fires.

His family tried to make bail for him, but because this was such a serious offense, bail was set at a high value. The family finally got a lawyer and he argued for lower bail.

However, by that time, we had arrested a second guard from the same agency for making some of the other fires.

Now the hotel was involved in suing the guard agency for damages. It's called a subrogation suit. In other words, the insurance company that pays off Helmsley is suing the insurance company of the security outfit, claiming they were negligent in screening these guys and putting them on the premises. The security firm's insurance company then replaces the family's lawyer with a very high-priced one, because if they can prove he is innocent they can beat the subrogation suit.

We surmised that the high-priced lawyer advised him to stay in jail because they didn't want him to get in trouble setting fires again.

The second one was at the Vista Hotel in the World Trade Center area. It was a similar case. In fact it was so similar it was eerie. The Vista was to open up that week, when they had a fire in one of the rooms. I arrested another guard.

You could just replace the names; the guards' backgrounds were the same: male Hispanics, early twenties, from single-parent homes, mother-dominated persons who had trouble with women, and with their absentee fathers. Both of them were heroes at the fires they set.

•

The family of the guard at the Vista was also going to make bail for him, but their lawyer advised them to wait until the disposition of the case. Again, we could only surmise that they didn't want him to set another fire in the interim.

Both of these fellows received the same sentence, two and a third to six years in prison.


I grew up in the Fort Greene housing project, what some people euphemistically call a ghetto, so working in the ghetto isn't as unique an experience to me as it is to some of the marshals who come from the suburbs.

The saddest case I had was in a ghetto. I saw four children die from two cents' worth of fire, one afternoon in the fall of 1981. I just couldn't believe it. It was in the Prospect Park section of Brooklyn, in a ground-floor apartment, right off the front stoop.

The mother had put the kids in for an afternoon nap. Her common-law husband repaired TV sets for a living. He was an unlicensed electrician.

Their building was abandoned by the landlord. It was in disrepair. The front door had warped, and it was scraping against the top sill. To close it you had to get behind it and slam it. Consequently, it was twice as hard to open. You had not only to coordinate the turning of the lock but you had to kick it open.

A television was left on. They had maybe 30 television sets in different stages of repair. Because there was no landlord to do the repairs on the building, this common-law husband started rewiring the house.

All of these elements came together, and they had a classical electrical-short right underneath the bedding, resulting in a fast fire which produced much smoke because of the foam rubber in the mattress.

Again, because of the disrepair of the building, they had

•

hung up plastic sheets on the inside of the windows to keep the heat in. Unfortunately, it made it hard for the smoke to get out.

The neighbors were a group of unemployed people who were hanging around at the time. They readily admitted that they were sharing a bottle. When they finally discovered the smoke they ran around and broke the window closest to where the fire had taken off. They thought they were doing good, but they really vented a small fire that was producing a lot of smoke, causing it to blow throughout the apartment.

A person who was inebriated tried to get into the front door, but he couldn't work the combination of opening the lock and kicking the door at the same time.

So, all of these factors came into play, and four children died.

We had to determine the cause and the point of origin of the fire. We discussed it with the DA but nothing ever came of it. The blame was so diffused. Who's really to blame? Is it society, or the individual who did his own repairs because they didn't have proper electricity, or the landlord who abandoned the building, or the people that damaged the door that made the landlord abandon the building? There did not seem to be any answers.

But, in the end there are 4 young children dead, in an afternoon, with adults less than 10 feet away. It just didn't make sense. It's sad because it was useless. It didn't have to happen. Anything that broke the chain of events that took place would have prevented it.

Everybody's had tragedy in their lives, but I wouldn't know what it is, within a short span of a few minutes, to lose your children. I don't know if I would have survived that. I mean the mother was able to give me a statement. I just was impressed with how stoic she was. If a tragedy like that happened in my family I think I'd be a bowl of jelly.

•

In April of 1987, Donald Green and I caught another fatal fire on Sutter Avenue, Box 1741, in Brooklyn. When the guys from Ladder 175 arrived, flames were coming out of several of the rear windows. The fire was confined to two bedrooms in an apartment on the second floor. When the fire fighters finally got into the apartment they found a woman naked and badly burned in the rear bedroom. She was down on all fours as if trying to protect her face.

In doing our physical we found evidence of pool-shaped burns on the wood floors, crossing the saddle of the doorway between both bedrooms. There were also signs of "fingering." We knew it was a low, rapid fire from the sharp demarcation between the burnt and unburnt areas, and because of the alligatoring of the paint on the walls. All of these signs pointed to a flammable-liquid fire.

We called for our forensic photo unit, the medical examiner, and the detectives from the 75th precinct, with whom we formed a partnership.

We started questioning people in and around the building.

In the ghetto there are what I call "two faces." What happens is there's peer pressure among the blacks, Hispanics and the whites you find in lower socioeconomic situations that produces the outside face. Which means, if you have a crowd of people around, nobody's going to step forward and give you any information. Nobody even wants to talk to you. In fact, they have this "you're on the other side of the fence" attitude.

It blows my mind to find out how cooperative the same people are a half hour later in their own home. I call it the inside face. ·

When they are standing in the crowd you don't bother them about cooperating. In higher crime areas they live in constant fear and can't be seen to be on your side. They're afraid of the drug addicts, the dealers, the burglars, and the murderers. All you have to have is one of these guys

·

on the block, and he'll strike fear into the rest of the street.

In the street you act rather standoffish and they do their role, until you're both alone.

We came to find out that the apartment belonged to the family of a 13-year-old girl called "Pudding." She told us that a guy by the name of Stanley Carver had been living in the apartment for the past few months. It seems that when Stanley first got into the apartment he chased out Pudding's mother and other family members, though Pudding continued to live there.

Pudding told us that she came home between two and three o'clock in the morning, and that it took Stanley several minutes to let her in, because the door was double locked, and since the door is broken you have to use a butter knife to open it. Once she was in the apartment, Stanley double locked the door again.

Stanley had a girl in his bedroom whom he introduced as Elvia. He then told Elvia that Pudding was his little sister.

Pudding stated that she didn't want to sleep in her room, because of what had happened to her boyfriend Blackie. He was shot and killed on the corner of Pennsylvania and Sutter the week before. So, Stanley took the mattress from her bed and put it on the floor of his room.

Pudding said that while she was trying to go to sleep, Stanley was having sex with Elvia doggy style. He was doing it really hard and Elvia's head was banging against the wall. Elvia screamed out in pain and asked Stanley to stop, but he wouldn't. Pudding told us that Stanley had an orange broom next to his bed, but when she got up later she didn't see it. Pudding did not know how Stanley was hurting Elvia, because out of respect for Stanley, she did not look. She just turned over and went to sleep.

The next thing she remembered was Stanley waking her up and telling her to get out because someone was trying to kill them, and there was a fire in the apartment.

She jumped up, grabbed her puppy, and ran barefoot

•

through her room and out of the apartment. All of the doors between the rooms and the door to the public hall were wide open. The only fire she saw was a small one coming out of the upper drawer of her dresser and that the stuffed animals, candy, and the can of deodorant on the top of the dresser were starting to burn. She said that Stanley went halfway down the stairs with her and then went back up.

She found a working phone and called the fire department. Then she went back to the front of the building. She then saw Stanley running up and down the stairs yelling, "Don't let her die! I know she is going to die, they are taking too long!"

Stanley then came into the street and got into a fight with one of the other tenants. After the cops broke up the fight, Stanley went to his father's house.

Detective Wuerth of the 75th PDU picked up Stanley at his father's place and brought him back for questioning.

Stanley told us how he was awakened by a popping and crackling noise in the other bedroom, even though the door between them was closed. He discovered that there was a fire. He tried to wake Elvia up, but she wouldn't get up. He then got Pudding up and ran out, but the flames were too much to get back into the apartment. He then gave us the details about his failed attempt to rescue Elvia from the outside.

There were too many inconsistencies. For example, we knew that if Pudding ran barefoot through the bedrooms there couldn't have been very much fire at that time. We also had statements from other witnesses that made us suspicious of Stanley's story.

I think that Stanley was actually trying to be a hero. I feel that he set the fire with the intent of rescuing Elvia, but the fire quickly got too big for him.

We arrested him on charges of arson and murder.

He spent three years on Riker's waiting to go to trial. During that time, he went through three different lawyers.

•

I can only guess that these lawyers probably tried to convince him to make some sort of plea bargain, but he refused.

His fourth lawyer was an 18B attorney. They are high-priced lawyers hired by the state to defend indigent prisoners. This program came into being because of a number of lawsuits asserting that these defendants were getting incompetent legal representation.

We recently went to trial on this case. Elvia's family was in the courtroom every day. They're very nice people, and they gave us a lot of support.

Stanley's lawyer based the defense on attacking our physical. The lawyer claimed that sheets from a canopy over Pudding's bed caught on fire and their falling onto the floor caused the burn patterns we observed.

However, Stanley had told us that he had taken down the canopy prior to the fire and thrown it in the backyard. The photo unit had a shot of the unburnt canopy lying in the backyard.

We were waiting outside the courtroom when the DA came out, shook our hands, and said, "The verdict is guilty." Elvia's family came out; they were very emotional. We had gotten close to them over the past few days.

Based on our testimony Stanley got convicted on arson one and Elvia's murder. The judge sentenced him to 2 consecutive 20-year terms in prison.

It gives you a high. After you have said good-bye and you're on the way home you start feeling it. You really know that you've made a difference. It comes on slow but it'll last longer than a fire high.


At a fire everything is a blur from the moment the alarm comes into the firehouse until you are finished overhauling.

I had the privilege of working with Captain Cerate and

•

Donald Butler in a South Bronx ladder company in the late sixties.

Cerate became one of the youngest battalion chiefs on the job. Donald Butler was one of the best fire fighters that I ever ran across.

I'm not being modest, but I was never as good as them. They used to talk to each other about what they were going to do, in a very calm manner, under these horrendous conditions. I'd be somewhere in the vicinity choking and gagging, trying to prevent myself from running out, but you had this overwhelming peer pressure to succeed.

I was only in the company a little while when we came upon one of these "worst case scenarios." We pulled up to a building with flames in some of the upper windows and this man hanging his two children out of one—all set to drop them to the ground.

The captain told me to take the roof.

I just ran inside and started up to the roof. All my training went out the window. I didn't know if I was supposed to take the stairs or use the fire escape. I was running so fast that I got a tremendous pain in my side. I thought I was having a heart attack, but I continued to the roof. I got the bulkhead door open and did what I had to do. It was critical to saving the people in the window.

Afterwards, the captain patted me on the back and said I did a great job. I don't know what chemicals my brain was releasing, but I got a real high. I was proud to be a fire fighter.

I remember I was exhausted, but I helped the engine take up the house. I felt as nothing could stop me.

In the marshals you get this quick type of high when you capture someone you have been chasing, but usually it takes a long time to get your emotional reward.

Last year I was under a lot of pressure, but I was on my way to register for school. I go to John Jay College of Criminal Justice on Fifty-ninth Street and Tenth Avenue.

•

My usual way of getting to school from my house is to park my car by the ferry terminal on Staten Island, take the ferry to Manhattan, and then take the subway uptown. Well, on that day last year there was a lot of construction going on around the ferry terminal and I couldn't find a parking space, so I pulled my car onto the boat. I got the first position in the middle slot and they filled up behind me. I got out of the car, locked it up, went upstairs, and had a cup of coffee. I was looking at what I was going to register for. I was very satisfied with myself. I was early and I would be at the front of the line for registration.

Next thing I knew, time flew by, and we were at the Manhattan side. I feel the ferry bump the slip and I jump off with the crowd. As usual I take the train up to school. I even stop in the cafeteria and have a cup of coffee with a few of my classmates. Then I go to the gym and stand on line waiting for the administration to open up registration. I'm in the beginning of the line. I meet this young lady that I'm friendly with. She said, "Hi, Bill. What are you doing after registration?"

"I'm heading home."

"Can you drop me off in Brooklyn?"

"Sure."

"Where can I meet you?"

"By my car."

"Where's that?"

I just paused, looked at her and said, "Where's my car? Holy shit, I left it on the Staten Island ferry!"

I jump up and flew to the phone. I had to call information to get the ferry terminal number. While I'm waiting for the phone to ring I'm practicing what I'm going to tell this guy when he comes on. Finally someone answers and I said, "I'm with the New York City fire department."

"And you left your car on the ferry, right."

"Yeah, that's me."

"Don't worry, we'll take care of your car, but you better contact your job."

•

"What's the matter?"

"Well, we saw your fire department identification in the windshield, but when the ferry docked there was nobody in the car so we notified the fire department. They think you've jumped overboard. So you better call your job."

I hung up and called my job. I get the BFI switchboard at fire department headquarters. It's ringing and ringing, but no one is answering. After about 10 rings, someone finally picks it up. I say, "I'd like to speak to Artie Crawford."

"You can't. There's an emergency going on and you can't get through. Call back later."

He's just about to hang up when I say, "No, no, I'm the emergency. I'm the emergency. I know what's going on! Put me through! Please, put me through."

"All the lines are tied up right now. I'll have to run to the back and give him the message."

I wait for him to get Artie up to the phone. I could hear everybody in the background. When I heard Artie pick up the phone I said, "Artie, it's me."

He said, "You all right?"

"Yeah."

Right away, just like a fireman, he said, "You fuckin' moron, you idiot, where the hell—what did you do?"

"I know. I left my car on the ferry."

"Don't you know we have fireboats out dragging the harbor for you? We think your body is in the river. The commissioner heard about it. He's over here now, and he wants to know what's going on."

Well, about an hour later I got down to the ferry terminal and I apologized to the dock master, "I'm sorry. If I remembered it earlier I would have ran right back."

He said, "No. Lucky you didn't come back, because behind you was an out-of-town funeral and they were going to kill you."

I remember that when I was locking up my car on the ferry other cars pulled up behind me. People got out of

•

71

them, and they were having a big argument. One guy said, "What the fuck are we doing on a ferry?"

The other guy said, "I wanted to show Grandma Manhattan by the boat."

The first guy said, "I've got to get down to Virginia."

Evidently it was a funeral procession that just came back from the graveyard in Staten Island, and half of them lived in the Bronx and they wanted to travel through Manhattan. The other half of the people wanted to leave right from Staten Island and go down south, and they were pissed off enough being on the ferry to begin with.

The dock master told me that one of the guys from the funeral came up and wanted to know what was going on. "Well, somebody parked their car, locked it, and left it on the ferry, and we have to wait for a tow truck to come and move the car."

"We don't need a fucking tow truck. I'll push the fuckin' car overboard. I've got a Cadillac." But they had to wait. I held them up for about 40 minutes until a tow truck came and removed my car. Not to mention that I also held up the ferry.

I had to go to the precinct and pick up my spare keys and my car. But in the meantime, there was a police officer assigned to the ferry boat, and he contacted the fire department.

The department said the guy who owns the car is a fire marshal and he has guns, so look in the trunk of his car and make sure he has no guns back there.

The officer put down the backseat and searched the back of the car and he found my gun-cleaning box. He also found some of our issued ammo. Our ammo is different from the NYPD. We have 110 gram, semi-jacketed, plus P .38 caliber bullets, which are more powerful than the standard police issued .38s. Evidently this young officer never saw this type of ammo before, and he thought they might be illegal dumdum bullets.

When I got to the precinct, at 4:00 in the afternoon, the

•

lieutenant was busy making roll call. I identified myself. The lieutenant called the officer assigned to the ferry to come out from the back. Then he said to the officer, "This is the marshal, give him his keys." However, the officer wasn't ready to give me my keys back; he wanted to question me about my ammo. He produced the ammo and said, "What's this?"

I said, "Well, that's our issue." I had some more in the pouch that I was carrying, and I showed it to him.

The officer interrupted the lieutenant's roll call and he said, "Lu, is this for real?"

The lieutenant said, "Yeah, yeah, now give him his keys and his ammunition and let him go."

Then something else happened and the officer interrupted the lieutenant again to ask him a question. By that time the lieutenant was steaming. The lieutenant, in front of everybody, said, "Okay, asshole, now give the fire marshal back his keys and his bullets and an apology and get him out of here."

When I showed up for work the next couple of tours there was a price to be paid. In my mail slot was a sardine wrapped in a napkin with a note reading "Billy sleeps with the fishes." I had Charlie the Tuna signs all over my locker and I was harassed. I'm still being harassed. The bottom line is that my claim to fame, after 21 years of working in many busy companies and doing lots of exciting things, is to be remembered as the guy who left his car on the ferry.

●

# 3

•

# FIRE MARSHAL

# JACK CARNEY

# (RET.)

In 1961, I was a laboratory medical technologist. It's a great job but not very exciting. At the insistence of some of my buddies I took the fire department test.

On March 27, 1965, I was appointed a New York City fire fighter. After probie school I was assigned to an engine company in lower Manhattan.

I could not believe that this was the fire department. I walked the floors all night long. I went to one or two small fires before I caught my first big all-hands.

We rolled in to a fire in an underground garage. Lines were stretched. Things didn't go well. People had disappeared. I was with a covering lieutenant. His name was John Daley. He came out of 103 Truck in Brooklyn. The next thing I knew it was just him and me on the line.

•

It was in a tunnel. The burning truck's tires blew up. To me it was like the end of the world. I was really scared. I thought that the whole place was blowing up. Lieutenant Daley said, "Take it easy, calm down."

After the fire, we went back to the firehouse. Daley lined everybody up. He then went down the line and asked everybody, "Where were you?"

I didn't know what was going on. Then he took me aside and said, "You get out of this company as fast as you can and get to one that is going to teach you how to fight fires."

I got a transfer to 219 Engine in Brooklyn. That's the borough I wanted to hang out in. The first night I was there Lieutenant Gene Greco said, "This is not the busiest firehouse in the City of New York, but it's the best." His statement has stuck with me to this day.

I constantly fought fires there for about eight months, but I would stand outside the firehouse, looking towards Brownsville, watching the black columns of smoke from the all-hands and the second-alarms coming in. It was even busier down in Brownsville.

Brownsville was where I wanted to go. Fire Fighter Pat Mahaney told me to go see Captain Eugene Ferran of 283 Engine in Brownsville.

I got assigned to 283 in 1966.

My second day tour in 283 was my first DOA. We roll out the door at 10:15 in the morning. Going down Sutter Avenue, we see the smoke coming out the windows.

I was a "mask man." We had these big heavy masks. The guys on the back step would stretch the line. Then I would run and open the Scott box, put the mask on, and come up the stairs.

As I came up to the top floor—it was a top-floor fire— the smoke was banked down. It was a bad fire. It was blowing out the windows. People were yelling, "There's a baby in there."

•

With that, Captain Ferran said, "Here's a mask, give him the line."

Well, I never before had a line in my hand for an apartment fire. I took the line and got so excited. The first thing I did was try to cool the place down. I listened to the instructions of everybody behind me. They kept saying, "Over your head." I directed the nozzle over my head, and I knock my helmet off. I couldn't tell them that the hot water is coming down and burning me, while they're saying, "Move in, move in."

I moved in as best I could and got to the front room. Naturally once you hit the front room the fire went out, to a degree. I was happy, but now two other "mask men" came and took the line away from me.

The baby was right there in front of us, lying across a trunk. It was three years old. The skin was split open.

It was my first real shock. I thought, if I only had the experience I could have done something about it.

It's 1:00 in the afternoon. I'm outside the firehouse, waiting to go home. I have tears in my eyes. I was really emotional about it. Big Eugene came out to me and said, "You did a great job out there."

I said, "Cap, come on, the first thing I did was knock my helmet off. By the way, I'm on medical leave—I'm burned." I had gauze all over my ears.

"There's nothing you could have done. That baby was dead when we rolled up."

He knew what I was thinking. It was amazing, but he'd been around a long time. He was a good officer. He made me feel a little better, but then after that I took to hate fire, or as we called it in Brownsville, "the devil himself." Whenever I went to a fire I would say, "You're not going to beat me."

I became the permanent nozzle man in 283. During the riots I went to as many as 25 calls, and maybe 2 or 3 all-hands, a night.

From 283 I got transfered across the apparatus floor to

•

Squad 4. That's when I had an incident with the neighbors. I was driving into work on July 4, 1970, when a child ran into the side of my car. I got out of the car in uniform, a T-shirt with the orange Squad 4 emblem, and reached over to pick up the child. The child was lying there screaming.

As I reached over I got slammed in the back of the head. The fist that hit me belonged to a guy with a gold tooth, a scar on his face, and black bushy hair. He was wearing a Black Panther uniform. He picked me up and slugged me again. I went against my car.

The people worked themselves into a frenzy. They surrounded me. One woman said, "We're going to kill you, you white motha. You don't give a shit."

I kept telling them, "Hey, wait a minute. I just made a rescue here two weeks ago. I was the fireman who rescued that man and the woman. I'm not like that—what are you talking about?"

The guy in the Panther uniform slapped me back and forth. Then he put me over the hood of my car and said, "We're going to tie you up. I got a rope."

One woman yells out, "I've seen that mother, man, he was driving 90 miles an hour. He didn't care about that child." All she did was incite them more.

Now they're tying me up, and they're telling me they're going to put me in a vacant building and set me on fire because I'm a fireman. That's how they're going to take care of me. I never felt more helpless in my entire life.

Unknown to me, an old woman, she was 70-something we later found out from the cops, pulled the alarm box. It was on Sutter and Hopkiss where this all took place.

Squad 4 made the turn from Sutter Avenue and saw the crowd. The officer got on the radio and said, "We have an accident." That's what it first appeared to be.

In the meantime, my car was being ripped apart. All the upholstery was being ripped and they were taking everything out of it.

I can't think of the fire fighter's name—he retired to

•

Ireland—but he saw me spread-eagle over the car. Then he yelled on the radio, "We need help. We need the police right away."

The men jumped off the truck with their tools. With that, Engine 283 came in. Then the locals started to scatter.

The Black Panther said, "I'm going to get you and your wife and your children." He had my address from my wallet. He gave me one more punch, and I rolled off the car.

That was my last tour in Squad 4. I was not allowed to work in Brownsville again. I was lifted out of there. Even today, I get very angry about it.

I had round-the-clock police protection at home for about two and a half months.

I grew up in a ghetto—Bedford-Stuyvesant. I'm an Irishman, and I'm a short Irishman—that's the worst kind. You always had reasons to fight, but now I was frightened to the point where I wanted to take out the Black Panther headquarters.

The fire department wanted to move me to Queens. I refused. I said, "You're punishing me for something I didn't do." I finally convinced Chief O'Hagen and he transferred me to 111 Truck in the Bedford-Stuyvesant section of Brooklyn. That was great, because 111 Truck had a great reputation.

The guy who beat me subsequently moved to California and joined the Symbionese Liberation Army, otherwise known as the SLA. They had kidnaped Patty Hearst, and the police were looking for her. Some members of the SLA had a shoot-out with the police. The house they were in caught on fire, from all the bullets and tear gas fired into it, and he burned to death. However, this was 9 or 10 years after he beat me up.

I was in 111 Truck only 8 months when I was assigned to Rescue 2.

Being in Rescue you were always in the midst of it. You're always surrounded by death. I have taken at least

•

11 New York City fire fighters out of buildings DOA. I have also taken out an unknown number of civilian DOAs, but I have saved quite a few.

During my time in Rescue 2 I was in three building collapses. They said there was a black cloud over me. I began to get the feeling that something was wrong.

I wanted to do something different. In 1978, I became a fire marshal. The reason I became a marshal was that I liked the forensic part of it—which is similar to being a lab technician, where I take a history, look at the slides, and then make an interpretation. I have to hunt. I have to search. There's always a step, step, step. I am organized. A "perfectionist" is what many people call me, but I don't think so. I am relentless, because the way I looked at it every patient was mine. Every fire I went to was mine. Every apartment, every room was mine. I take possession whether with a microscope, or a nozzle, or with authority as a fire marshal. It's taking control. This is my domain and I want to do this right.

The Redcap unit was formed two years after I joined the marshals, but I was in Red Cap from day one. To me it was the answer to stopping the fire fighters from moving on those trucks for false alarms, because I resented it strongly as a fireman in Brownsville.

We had 10,401 runs in Squad 4 in 1 year. In a typical 24-hour period you would have 40 runs. You would do 3 fires, a couple of shootings, and a couple of knifings, but 25 of the 40 would be false alarms. I figured as a marshal I could do something about that.

I also felt that if you go in there and let the people know this is going to be looked at, the fires and false alarms would stop to a degree. It has been proven with the Red Cap.

Red Cap was probably the best time. It was also the roughest time, so to speak.

We responded to every box that came in, within our zone. We would either beat the apparatus in or come in

•

alongside them. They would say, "What are you guys doing here?"

When the trucks left after a false alarm, we'd start talking to people. "Gee, what happened?"

"I don't know man, they always come—"

"Do you think it's broken?"

"Man, this guy down the block pulled it."

"Really . . . did you see him?"

"Yeah."

"Oh, that's great. Which guy?"

"Man, I ain't gonna tell you."

"Why not?"

"Because, man that's . . ."

"Do you want that fire truck here when somebody else may be getting killed over in the next block, or a kid gets killed, or a fire truck runs into a car? I know you don't want that."

"You're right, man, 'cause every time they comes we end up with, you know, being harassed later on by the police."

"Well, I'm not a cop."

"You carry a gun."

"I know, but it's only for my protection, not for anybody else. It's only when I'm caught in the middle of something." They understood that.

"I'm still a fireman."

"Eventually they would tell me who the person was who pulled the alarm. Then I would go up to that person and say, "Where is the fire?"

"What do you mean?"

"Well, if there was no fire, why did you pull the alarm?"

"I didn't mean to, you know."

"Okay, now you're under arrest." We used to lock them up left and right for false alarms.

Sometimes, a good guy named Emile Harnesfager and I did stakeouts on heavy false-alarm boxes that came in

•

every half hour. We used binoculars while sitting in a marked car two blocks away.

We would radio the dispatcher, "Dispatcher, this is Squad 41 Alpha. We are sitting on Box 2635. If it comes in, hit on it right away, and let us know—we're doing a stakeout."

"Okay, Squad 41 Alpha."

I was always the driver. Emile handled the binoculars. He'd say, "He's hitting it, Jack. He's hitting it."

"Are you sure?"

The dispatcher radioed, "Squad 41 Alpha, it's coming in now."

"Ten-four. Have the units continue in." You have to have a fire unit on the scene to say it's a false alarm.

I'm fortunate, or smart enough, to say that most of my cases have confessions. These people love to talk to you.

They do much of this stuff because they're just angry. For example, a pregnant gal got gasoline from a gas station and went back to an apartment on a second floor of this tenement. She took bags of garbage with her from the first floor and placed them right against the door to the boy-friend's apartment. She poured the gasoline over the bags and lit a match.

The guy in the fire apartment was alone, but she thought she heard him making love to another girl. They had broken up, but she's pregnant with his child.

He said, "Man, it had to be my girl."

Somebody else said they saw a pregnant girl in a red coat coming out of the building at the time of the fire.

We drive around the streets and we find her. She's 16 years old, skinny, 4 feet 11 inches, a small girl. Her pregnancy was bigger than her height.

We took her in the car and asked her, "Where were you tonight? Were you in your boyfriend's building?"

"Not me."

"Okay. That's lie number one. That's okay. You can tell me anything you want but that's number one."

•

She looked at me and said, "Okay."

"Were you walking by the building? That's what you were doing. That's why people saw you. You were just walking by."

"Yeah, that's right. I only walked by the building."

Once she gave me that information, that was it. She was on the scene as far as I'm concerned. Then we work it out further. I asked, "You and your boyfriend having trouble? You're not going with him anymore, or what?"

"He kicked me out."

"He's got a new woman?"

"Yeah."

I said, "Well, yeah, I'd do the same thing if I were you. The hell with him. If I can't have him nobody else is going to get him. I'll cook his ass."

She said, "Yeah, I would. I did."

When I interview somebody I write down exactly what they're telling me word for word. That was probably why I had more convictions than anybody else, because my notes would not be hieroglyphics or a one-liner. If I talk to you I automatically describe everything in detail, including the time, date, and a description of the surroundings.

It's also important how you make the people feel. You have to become one of them and be their friend.

I used to tell all my partners, "Look, if you come in and you slap people around maybe you'll feel good, but you'll never get anywhere. That's the end of it. It's not like in the movies."

I always got my man or woman, but I always did it with a little finesse and a very quiet demeanor.

I rarely yelled at anybody. Maybe my partners would yell. That's how the part of the movie where you see the cops playing "good buy–bad guy" works.

Ed Peknic was one of the best partners, as far as playing the bad guy, because he's a great actor, and he is an ex-cop besides. He is also taller than me and ugly.

•

He'd say, "Hey Jack, this bitch is lying to you. I'm going out to get a soda. If she doesn't tell you anything by the time I get back, I don't care what you do, but I'm locking her up."

Well, that's what I was going to do, but I first want her to tell me why she did it. That was most important to me. Not so much how, I figured that out—it was gasoline. I was not satisfied with a confession if the person didn't tell me the motivation behind it.

I think it gave the case a sense of completeness. So that when I take their liberty away the fact that they told me why means that they knew they did wrong, unless they were complete idiots.

I felt sorry for that pregnant girl. She really felt alone. She sat in our car and cried and cried. We got her soda. We got her coffee. She needed help. If I could have done something for her, without arresting her, I would have. There's no question in my mind about that. Most of the people that I've arrested do not realize when they set a fire how dangerous it is.

To let other people know that you can't set fires, I always made sure that the person being arrested was going out very slowly, in cuffs, unless there's a hostile crowd, or it was a kid or a person with emotional problems.

One Sunday morning, I went out to Bay Ridge with another fire marshal to investigate a fire in a bedroom of an apartment. The fire was incendiary.

The husband is there. He's 74 years old. His wife is a stout, 62-year-old Italian woman. She was cooking sauce.

We question the husband. He says, "She set the fire. She thinks I'm going out with some girl."

Then I find out that she's very menopausal. She's depressed. I said to my partner, "Hey, she did the fire."

"No question about it."

She says, "Yeah, I burn his clothes."

My partner puts one cuff on her. She grabbed his gun.

•

We do a tarantella for 15 minutes with her hand on his gun. She wanted to shoot the husband.

After we got her calmed down we didn't want to parade her out in cuffs. So we front cuffed her and took her out with her coat over her hands.


I used to love it. I used to live for the next tour. The Red Cap to me was probably the most successful anti-arson unit in the country. What made it work was presence and the people knowing that somebody cares out there.

We had this Red Cap operation going in Brooklyn. There was an occupied building that had many fires in it. We're talking about 10 to 12 fires within a 2-month period. Everybody was hopped up about it.

It was a 4-story multiple dwelling somewhere on Sixth Avenue and around Forty-fourth Street in Brooklyn. There were four occupied apartments still in the building, but there was only one vacant apartment that was not burned. Not major fires, one-roomers; kerosene or gasoline was used.

We had a suspect. We felt it was the superintendent of the building. He was trying to get the tenants out for the landlord. Anyway, he was an ornery guy.

We decided to do a surveillance in the building, on the four-to-two shift. Bobby Brown and I were going to do the stakeout in the vacant apartment with a backup unit a block away with radios.

We had to get in the building undetected, but there were people hanging around the entrance, so we went into the building on the opposite end of the corner. The idea was to go up to the roof, across the buildings, and down the fire escape to the vacant apartment. Great plan.

We go up four flights of stairs. We get to the top landing and all of a sudden the door of an apartment opens. This old man sees us. Right away we flip out our shields and say, "Police."

•

The guy says, "No, no, no. Dog up on the roof."

I said, "I love dogs. No problem." But my partner Bobby Brown, who is black, turned white. On the staircase there's a six-foot long two-by-four. Bobby says, "Yeah, I'll take this."

I said, "Don't worry about it. Come on, we got to get in place."

We go up, open the bulkhead door, and step out onto the roof. As we walk around the bulkhead we see a big, black Doberman pinscher up on the parapet looking at the street. He's the guard dog for the roof.

Bobby goes, "Holy Christ."

The dog turns around and spots us the same time as we see him. I said, "It's okay, puppy, relax. Nice doggy, nice doggy. Bobby, come on, let's go." As we take a move the dog comes running towards us.

Well, Bobby doesn't know what to do with the two-by-four, so he drops it. I trip over it. We both land on the ground, waiting to be eaten alive. The dog jumps right over our head and into the air shaft. It falls four stories, but lands on a pile of garbage. We were hysterical. I couldn't believe it. We're saying, this is some covert operation.

It's dark, but we don't want to shine lights around. However, we go over the rooftops and down the back fire escape to the third-floor vacant apartment.

Underneath us is an already burned-out apartment.

I sit against the wall under the window of the fire escape. Bobby goes in the next room, but the wall's all broken out, so we can see one another. We both can see the hallway and the door, though it's pitch black. Both of us have our weapons drawn. We both have flashlights laying by our feet. We settle in for the night. I turn on the radio and say, "That's it, no more talk. We're set."

An hour and a half goes by when Bob says, "I smell something."

"Oh man, yeah."

•

In less than five seconds the rooms start filling up with smoke. Flames come roaring out the back windows.

We can't go down the fire escape. We run to the door and open it. The fire is out of the vacant apartment below us and up the stairs. We have two trapped fire marshals.

We call on the radio, "Mayday. Mayday. Ten-75, get firemen here." The unit in the street can't come up to help us.

The guy must have known we were here, so he poured out two gallons of gasoline and sets the fire underneath us.

We can't go down the fire escape. We can't go up. We had to go into the public hall, on our bellies, and crawl. We kick in the door to the front apartment and get out to the front window.

An aerial ladder came and got us off.

That was some act. I couldn't believe it. I wanted to kill the guy.

The superintendent was arrested, not for arson, but for assault on a police officer. To this day I swear he did it, because the fires stopped after he got locked up.


I like doing searches in vacant buildings. I like catching guys. "Hey, look, I saw you here once before. This is the second time."

"I was just taking a piss."

"Really, you had to go all the way up to the second floor to take it? What's this? Drugs?"

I would break the vials, break the needles, and throw the junk on the floor.

"Don't just throw it out, at least use it yourself," the junkies would tell me.

They were the ones that were setting the fires. They would build a little fire to cook their drugs. Then they'd get high and wouldn't put it out. Two hours later we have a 10-75 in a vacant building.

•

"Hey, this building belongs to me, and you're under arrest." Now the people know that you can't go into a vacant building.

I have three sons who are police officers, and I love the police department, but they wouldn't go in a vacant building for a million dollars. They say, "That's crazy."

Fire marshals go into them all the time. We aren't afraid of holes in the floor. We aren't afraid of burned-out rooms. We aren't afraid of missing steps. We frequently encountered these conditions when we fought fires.

On a hot summer night in July of '82, I had a brand new fire marshal, Bobby Mauro, working with me.

He has no hair, dark complexion, mustache. Everybody thinks he's black, but he's Italian.

We responded to a scene of a 10-41, Code 1, a suspicious fire in a vacant apartment, on Union Street between Third and Fourth avenues in Brooklyn. We came in after the fact. The small fire was out, and the fire engines had already left.

As we get out of the car there was a very agitated woman up on the fourth floor of the fire building yelling out the window in Spanish.

I said, "Bob, number one—be aware of your surroundings."

He said, "Okay, Jack."

I opened my jacket. I tell him to open his jacket.

We're walking west on Union Street towards the fire building, when out comes this strange-looking dude: A Spanish guy, fuzzy hair, no shirt, wearing a pair of pants with suspenders. He runs across the street into another building.

Bob and I stop. We're about 100 feet from the fire building. Suddenly, these other people, who are hanging out of the windows in the building, also start screaming.

I said to Bobby, "That's probably our fire setter."

As we walk to the stoop a woman comes running up to me and says, "I know who did the fire." She's speaking

•

in Spanish and broken English. "He's loco. He's going to get a gun and kill people."

I said, "Whoa, whoa, calm down."

As she's talking to me she looks across the street to the building where this man ran into. She says, "Oh, *mira, mira, mira.*"

I look and he's coming out across the street. I yell, "Bobby, watch this guy!"

I always tell my partners that they are not here to get involved in the interview. If I'm doing the case, I'm doing the interview. All I want them to do is look around and make sure I'm protected. And if there's something wrong, call me by my first name. Then I'll know something's going down. The same holds if I call them by their first name.

I don't know what this guy's got. He has his hand over a brown handle of something in his waistband.

As he comes towards us he doesn't give us any notice at all. The man is wild and crazy-looking.

I do a glade, which is a sidestepping of my right foot, and reach back for my Colt two-inch Detective Special. Usually you don't have to show your gun. You just put your coattail back and that gesture alone stops many problems.

I said, "Excuse me. Can I talk to you a minute?"

I'm raising my shield in front of this man, but he doesn't even look at me. He's 10 feet away, and he's walking towards me. "Excuse me!"

Bob reaches for his little two-inch Detective Special, and he raises his shield. He's on the opposite side of me. The man has to walk between the both of us.

Bob says, "Hey, man stop. Police officers, we want to talk to you."

"Fuck you."

He goes up the steps.

"Bobby, watch this guy."

Now the adrenaline starts. The shields stay out. You just flip the leather holder over and put it in your pocket.

•

The guy is now halfway up the steps.

I draw my weapon and yell, "police." Bobby draws his weapon. We have a bead on him.

He turns towards us with this thing in his waistband.

I'm screaming, "Drop it! Get your hand away from your waistband!"

"Fuck you."

He turns his back on us. As he walks into the building the woman who is up in the fourth-floor window yells, "Oh *mira, mira.* He's coming up to kill me. He's going to kill me."

We catch up with him on the third floor. The hallway is about 35 feet long. It's very narrow. Bob and I cannot stand next to each other. Bob's a pretty big guy, so he stands to one side behind me. We both have our guns pointing down the hall.

I yell, "Stop! Stop!"

He won't turn around.

With that, I do something that's legal, but not smart. I pulled the hammer back.

He hears the two clicks of a Colt.

"Okay, man, next one you're hit."

He stops.

I think, "Oh, thank God." Then I say, "Turn around."

He turns around, looks at us, and says, "Fuck you, *maricon.*"

Two guns are pointing at this man, and he still has his hand on his waist. I'm feeling resentment that he would not obey my commands immediately. I am also waiting for a gun to come out, any second now.

I'm ready to pull the trigger, but I'm telling Bob, "don't shoot, don't shoot," even though we had enough cause to shoot him right then and there. We were in fear of our lives, and we thought he had a gun.

We start going down the hall. He backs up to the door of the apartment behind him. In front and to his right is a staircase that also goes down to the second floor. As he

•

backs up to the door, he takes the object out of his waist-band and raises it over his head. It's a meat cleaver—a big, shiny meat cleaver.

We stop dead in our tracks. I said to myself, He can throw that and hit one of us, no question about it. I go into a combat crouch. I tell him, "Fuck you. That's it. You're dead."

He stops, puts his hands up, and drops the meat cleaver.

I was feeling pissed off that he made me do all this. I was really angry at this guy and I said, "You son of a bitch."

I was afraid Bob was prematurely going to shoot, because he was sweating. I mean, we were both scared. But he didn't.

We went up to the guy. The meat cleaver is now on the ground. Mistake number one—we didn't kick the meat cleaver away. Bob puts his gun away. I keep my gun out.

"You're under arrest for menacing." We figure the hell with the arson. Once we have him cuffed he'll tell us he did the fire anyway.

Bob tells him to turn around. The guy doesn't turn around. Bob again says, "turn around." Bob is reaching for his cuffs, but the guy won't turn around.

Bob has giant gorilla hands. Bobby's fist is bigger than my two fists.

He flips the guy around and pushes his head into the door. The guy turns right back around. Bobby says, "Hey man, what are you, stupid?" Bobby is cursing. I'm cursing. Bobby turns him around again, but he comes right back.

Bobby lets him have a shot in the stomach. It's allowed— he's resisting arrest.

The guy doubles over, and goes down on his knees. He's yelling, "Aieeee"—that Spanish yell which sounds horrendous in the hallway. Unknown to Bobby and me, the guy reaches for the meat cleaver.

Bobby says, "Holy shit!"

The guy comes up and around and punches my gun

•

into my face with his left hand. I'm thrown against the wall. With his right hand he raises the meat cleaver up over his head.

Bobby quickly draws and fires. Oh, great. It made a tremendous noise because of the marble staircase.

When Bobby turned to shoot the guy, his back was at the barrel of my gun. The City gave me a medal for not shooting my partner in that quick one-second incident. I often think of why I didn't pull the trigger, because it's a reflex particularly with the noise of another gun going off.

Bobby grabs him and they wrestle. The man breaks away. Bobby falls down the flight of stairs, firing two more rounds at this guy.

I get to the top of the stairs. I'm on the radio screaming for backup, "Shots fired, marshals involved." Thank God they knew where I was. I always had a black cloud over me, so every time I went to a building I used to say to the dispatcher. This is 41 Alpha. We're going in at such and such location." Then if I scream, "41 Alpha is in trouble," I don't have to worry about an address.

The guy went down the stairs and through the aerie-way, where the garbage cans are kept, to the front of the building.

Bobby is lying at the bottom of the stairs. I said, "Bobby, I'm going to the front of the building, I'm going to cut him off."

"Get the son of a bitch."

I run with my gun pointed up in the air. I beat him to the front. He comes crawling out on his hands and knees.

With my gun pointed at him I say, "Stop. Give it up, man."

He's bleeding. His leg is shattered, but he keeps crawling another 10 to 15 feet. I'm walking along with him. I'm thinking, "This is stupid."

He's yelling something in Spanish. To this day, I wish I spoke Spanish, because he's telling them something that I think is not true, but I don't know what he's saying.

•

The drums are starting to beat. People are screaming. People are opening windows. People are coming out of buildings.

I step on his bad leg. He lets out a scream and then he stops. I put my gun away and cuff him quickly. He's now off the sidewalk and in the street.

In the meantime, a woman comes running over. She is a small, heavyset Spanish woman.

He is again screaming everything in the world.

I take my gun back out after I cuff him. Bobby has not come out of the building yet. I figured he's hurt, but he is probably all right.

The woman rips open her blouse and falls on top of my prisoner face up. Her breasts are 44 double D's. Her tits are up in the sky. And people are looking as if I shot her.

Now the crowd becomes hostile. People start pushing me away from my prisoner. I said, "Whoa, stay back, stay back. Police, police, stay back."

They wouldn't listen. They thought I shot her. They thought I shot him in the back, while he was cuffed. By this time they come running from all over.

Bobby comes out and says, "You got him. Fuck him."

"Bobby, look at this crowd."

The crowd starts coming at us, pushing us away from our prisoner. Now we don't see our prisoner. We don't see the woman anymore. We're surrounded by about 100 to 120 people. We are in deep shit. I'm screaming on the radio, "Hey, get the world here! We're in trouble. We got a hostile crowd situation."

Another woman runs up to me, "You think you're smart. You shoot people. You don't give a fuck. You're a rat's ass."

I said, "Lady, step back, get away from me."

"Go ahead, shoot me! Shoot me," she yells. Then she comes up to my gun and pushes her tit right into it.

A dog attacks Bobby. Bobby is kicking at a dog.

I said, "I'm going to fire. Stop, get away."

•

The cops came in first. I'll never forget this one uni-formed older cop running through the crowd. He's punch-ing people out of the way, because we're yelling, "police," and we're showing the shield so they know we are the good guys. He comes up behind me, pats me on the left cheek of my ass, and says, "It's okay, son. You're safe, just ease up on the trigger." My finger was white. "You can keep it, but just ease up on the trigger."

I said, "Oh man, how am I going to explain this."

Then John Stickevers comes rolling in. I was working for him that night. He throws us in the car and says, "Get these men out of here right away."

They took our prisoner. He was shot once in the side and twice in the leg. Bobby did very well.

The guy ended up with a limp. Three years later, he was shot and killed by the police.

We're at the precinct. Talk about stress. A deputy in-spector from the police department comes over and asks, "How are you doing? You all right?"

I said, "Yeah, yeah."

He says, "Fellows, can I have a look at your weapons?"

We pull out our weapons.

"Fire marshal, you reloaded this?"

I said, "No, I didn't fire."

"Oh, you didn't fire."

Then he opens up Bobby's gun. "How many rounds did you fire?"

"One."

The inspector said, "Well, I got three shells missing." He then looks at me and says, "He fired three times?"

I don't know what to say.

Bobby says, "Did I? I'm not sure."

Then they had a hearing with seven or eight witnesses that claimed the man with the gray hair shot Julio in the back, while he was handcuffed in the street. Fortunately, Bobby has no hair; thank God there were some good cit-izens that told how it really happened.

•

The next day, the ballistics guy comes to me and says, "Can I see your gun?"

I said, "Sure," for the fourth time. "I'm showing it to everybody."

He said, "You fucking detectives," just like that. "You guys like to carry these two-inch guns and look sexy like in the movies. They'll never get you out of trouble. If you had a real gun that man wouldn't get up. He wouldn't be able to hurt you."

"What's a real gun?"

"Four inch." He didn't even say what kind.

The following day I went to a gun shop and said, "Give me the ugliest, meanest-looking gun you have." I have that gun to this day. That four inch has been my trademark with people. The four inch, silver Colt Python .357 Magnum calms people down. It's not the same kind of weapon as the police have, though I only loaded it with our +P .38 caliber ammo.

There were people who knew my gun from a distance of 30 feet. When you drew that gun in the ghetto they knew you were not NYPD. The world stopped. The people would actually say, "Man, you own this hallway."


We did an investigation of a fire on Third Avenue and Fortieth Street in the Sunset Park section of Brooklyn. The fire was an all-hands with three people jumping out windows. A young woman broke her pelvis trying to escape the flames. It was a really bad fire.

We end up locking up a 12-year-old boy. He set fire to the vestibule. He did it because his mother locked him out.

Bill Mulhall and I had to take the kid to the Spofford Juvenile Detention Center. On the way we stopped at a Burger King and got him a hamburger.

I always did that for all my prisoners. "I'm taking you in. If you're hungry tell me now, because once you get to the precinct I sort of lose control." I know it's strange, but

•

that's the way I am. I got you. The hunt is over. You are no longer an adversary to me.

We came back from Spofford around 2:00 in the morning. Bill and I are making out our reports in an old house trailer parked on the sidewalk of Fourth Avenue.

We hear a loud noise like a shot. I look at Bill and I said, "What the hell was that?"

"I don't know."

Then we hear loud yelling and screaming. As I run to the front door of the trailer I tell the phone man, "Dial 911, get the cops. There's some problem outside."

I open the door and I see two cars stopped in the middle of the street, a crowd of 8 to 10 people. Then I see a man, and he has a silver object in his hand. He fires a round into the ground.

I tell the phone man, "Tell them shots fired. Marshals on the scene. Plainclothes police officers present. Be careful with their weapons."

Bill and I run out. I say to him, "Shields out. I don't want to get shot." My biggest fear in my 10 years of being a marshal is being shot by another police officer, because we normally wore scruffy clothes, sneakers, and carried odd guns, though we never had long hair.

We move alongside a bunch of cars until we get about 75 feet from the guy with the gun. He is in the middle of Fourth Avenue. Billy and I are at an angle to him, but we're behind a parked car. I yell, "Police! Drop your gun. Drop the gun."

He's not even looking at me. Now he has the gun pointed at the crowd.

Somebody to my right yells, "He don't understand you man, he's Spanish."

"Then you tell him in Spanish to drop the gun, or I'm going to kill him. Tell him I'm going to shoot him."

The gunman still doesn't respond. Now I stand straight up, and I get a bead on him.

He starts turning towards me with the gun.

•

"Oh, man." I'm aiming at his head, because it's the biggest target I have. As I'm pulling the trigger, he turns back to the crowd, gets into a combat crouch, and fires. He hits someone. I fire at the same time. I hit him and he goes down.

The crowd goes wild. I don't know what's going on. My partner can't believe it.

Some people started to jump into one of the cars in the middle of the street. I point my gun at the front wheel of the car and blow out the tire. They drive away with a blown-out tire. I said, "Something's dirty here."

I then tell Bill to cover me. I come down on my suspect faceup lying in the street, with his head pointing towards me, with my weapon poised in case he makes a move.

The gun is still in his hand. I tell Bill, "I'm going to take the gun." We hear people screaming. I run up to him, step on his hand and take the gun.

I roll him over and there is blood all over his back. I stopped for a second and then I rear cuffed him. He's making noises. I said, "Oh, he's not dead."

Bill starts to come over to me. However, some woman grabs him and starts beating on him. He slaps her down with his gun.

The next thing I know I have four guys running towards me, screaming in Spanish. I draw down on them and say, "That's it, down on the ground. Everybody down." They understood someone pointing a gun at them and then pointing at the ground, even if they didn't understand English.

I have no more cuffs. Bill has a woman, almost unconscious, behind us. We now have a crowd of about 25 to 30.

The cops come rolling in like gangbusters. This blond police officer, who's first name is Pat, yells, "Look out for the guy with the gray hair and the big gun!"

I don't want to turn around, but my shield is in front

of me and my back is to the cop. I yell, "Fire marshal! I'm a fire marshal!"

"Show me some identification."

I just held up the shield, without turning towards her, and I said, "I need some help."

The cops were fantastic. We cuffed the people on the ground. The shooter went to the hospital. They took me to the precinct.

I had everybody coming over to me asking, "Are you all right?"

"I'm fine."

They wanted to send me for counseling, because they didn't like my reaction. A department lawyer comes walking over to me in the precinct, about two hours after the incident, and he says, "When is the last time you had a drink?"

I said, "What? What are you talking about?"

"When is the last time you had a drink, Fire Marshal?"

"I'm not answering your question."

He says, "I'm the attorney for the fire marshals' office."

"I don't care who you are. I'm not answering your question."

Then Chief Fire Marshal John Regan comes over to me and he says, "Jack, I want you to take it easy. Relax, tell your story."

"Chief, I already got two yellow sheets of my story written down. I have no problem with this. The man had a gun. He's going to shoot people. As a matter of fact, he *had* shot somebody. I took him out."

Then Deputy Chief Tony Romero comes over and he says, "Jack, you're emotional."

"Tony, I'm not emotional. Just give me some paper. I need some more paper so I can write the whole story out."

The first report from the hospital came back that the man was shot twice in the back. A detective runs over to me and asks, "Did you put two rounds in his back?"

•

"I don't think so. I shot once at him."

What happened was, when I was about to shoot at his head he turned and went down into a combat crouch. I brought my gun down too far and caught him in his right side, above the shoulder blade. The bullet hit the spinal cord, fragmented into two pieces, and came out his back. One entrance wound, two exit wounds.

I hit his spinal column. That's why he went down so fast and never moved. He was paralyzed, but who knew then.

We just finished the civil court case last year, which was five years after the shooting. He walked away with $150,000. It didn't matter that he was shooting. He's paralyzed for the rest of his life.

They all sue the City of New York if they're shot by a cop, because maybe I didn't have the proper training, or maybe I used the wrong gun, or maybe I was drinking, or maybe I didn't give fair warning.

If you get killed, however, your wife can't sue. It's in the line of duty.

This whole incident was the result of a fender bender. One car hit the rear bumper of another. The man with the gun was in the lead car. He got bumped again. He stopped the car in the middle of Fourth Avenue and got out. The people in the second car also got out. Both parties had been drinking. They started arguing. He drew a .32 caliber pistol and fired one round into the air. That's the round we heard in the van.

Other cars, with friends of the people in the second car, stopped. Then the neighborhood people started to come out.

The shooter fired again, but into the ground. With the third shot he hit a guy in the chest.

The wounded guy was part of the crew in the second car, but he didn't come forward. No one knew he was shot.

I said, "He shot into the crowd."

•

"Did he hit anybody?"

"I guess not. Nobody fell."

The next day, a young Irish detective with a brogue checked the hospitals out on his own and found this guy.

They never even went to the grand jury with this shooting. I had 40 witnesses. That's how clean it was.

I'm against touching a prisoner. My reputation is that in 10 years I have never touched a prisoner except for 2 times in my life.

The first incident where I actually hit somebody was when I was after a 16-year-old Irish kid. His name was Billy. He and another fellow turned a gas stove on but blew out the flames. Then they lit two punks in the apartment that they had just burglarized. They figured if they set the apartment on fire or blew it up, they wouldn't have their fingerprints all over the place.

To make a long story short, we found out who the kid was. We had a good, solid case. We went to the building where he lived. His mother answered the door. They lived in a basement apartment, with windows opening onto a courtyard.

I said, "How you doing? Is Billy here?"

"No, he's not here."

As she's saying that, I see this young boy run into view, going from one room to another. I said, "Excuse me, lady." And I push my way in. Bobby Mauro is with me again.

I run into this room, which is a big bedroom/living room combination. It has a couch and a bed in it.

I said, "Where did he go? Bobby, go into the courtyard—he went out the window."

In case he didn't I start searching behind the couch and under the bed. Then I walk into this closet. It's five feet wide by eight feet deep, with clothes racks all around the room. I look on the ground for his feet. I don't see them, so I walk out of the closet.

Bobby comes to the window. He's outside looking in. He says, "He can't get out, there's a stone wall here."

•

"Well, then I'll check the closet again."

I go back into the closet and start moving clothes. Then this blur of clothing comes right over my head and this kid starts taking a rap at me. I push some of the clothes off and put them over him. Now we both have clothes over us.

I just started to punch for my life. I couldn't stop punching.

The kid's screaming, "Help, help, help."

Bobby hears me screaming, "You son of a bitch!" I thought I was dead. I didn't know what it was.

Well, the mother come in. Then Bobby runs in. We take the clothes off this guy. He's 105 or 110 pounds. He's a skinny little Irish kid, who if he hit me full blast in the face, it wouldn't hurt at all.

I smacked him for making me punch him.

"You stupid. . . , I could have shot you." If I had my gun out I truly would have fired. Thank God I didn't have the gun out. I don't think I initially felt that threatened, but what happened scared the shit out of me.

We arrested him. I found out later that he had been helped by a New York City Police officer after getting into trouble before, but there was a warrant on him for stealing the officer's gun. It was a sad case.

In the other case a man from Jamaica burned cardboard in the staircase of his father's home in Bushwick. His father was a God-fearing man.

Eddie Peknic was my partner.

We responded to the job. The father says, "My son Charles did the fire."

"Why did he do the fire?"

"We had a fight. I threw him out. He's mad at me. He's crazy. He's got a gun."

It's a bright, sunny day. We search for Charles in the neighborhood, but we can't find him. We tell the father, "If he comes here, call this number and tell them that Charles is here. We'll know what to do."

•

A little while later we get a call that he is at the father's house. We respond up to Bushwick. He heard the father call; by the time we get there Charles had run away.

The father calls again. He tells us that his son is now outside the house. This time we go the wrong way up a one-way street, just like you see on TV, Charles is leaning against an iron fence with a coat over his right arm.

We jump out of the car and yell, "Charles, put your hands up!"

"What for?"

We have our hands on our weapons. "Get your hands away from your body."

"What for, man?"

That's it—both Eddie and I draw our weapons and go into a combat crouch. Eddie always goes to my left. Eddie and I are like one. If he makes the initial move, I drop back to his right and go for cover. It's a team concept.

"Drop the gun. Get your hands away from your body." We're inching up close to him.

"Why?"

"Get your hands away, or I'm going to blow you away."

Charles finally moves his hand away from his body. I thought for sure he's going to shoot us, but he had no weapon. We told him to drop the coat.

I said, "You're under arrest." I figure this is it—we'll just cuff him.

Eddie's got his gun out. I put my gun away. I then reach for Charles's right hand, but he pulls it away. I say, "Hey man."

Charles looks me in the face and says, "I don't care what you do to me, you're not going to put cuffs on me. Fuck you."

I bend his arm. He starts fighting. We fall down. I'm wrestling with him. With that, Eddie kneels on his back, puts his gun to the back of the guy's head and says, "Stop it man. Just give him your arm."

"Shoot me, I don't care."

•

It's a no-win situation. It has always been said if you don't want to be cuffed, 10 men can't do it. Well, we tried to do it.

We had to put our guns away. What are we going to do? Hit him with the gun? Even that wouldn't work. The guy was beside himself.

We were at a lamppost, and he was hanging on to it. I have him in a headlock. He knocks me down to the ground. We're under a car. We're over a car. This went on for 10 to 12 minutes. We can't even talk on the radio—our hands are full. We're afraid this guy's going to grab one of our guns or punch the shit out of us, because he's strong.

Eddie is fighting for his life. Eddie is going to say, "Jack messed up again."

I'm saying, "Help me, goddamn it."

"I am helping you."

We start to laugh. I don't know why. I said, "Will you stop it?"

The next thing we know a cabdriver pulls up. His arm is out the window. We have Charles's body over the hood of a car facing into the street, though his feet are still on the sidewalk. The cabdriver is looking right at him.

Ed and I are sweating and exhausted.

The cabbie is a black man. He has at least 18-inch biceps. He says, "Hey, you need any help?"

"Hunh?"

He pulls out a silver police shield. "I'm driving a cab—off duty."

I said, "Yeah."

He gets out of the cab. He is almost as big as it is. He comes behind Charles and says, "Give me the other hand." End of story.

In the car going to the precinct I'm starting to feel pain all over my body. I get mad. It is the first time that I almost punched a handcuffed man, but I don't.

•

In 1984 I was working with Al Post. We went to investigate a fire in Williamsburg, Brooklyn. It was a time in Williamsburg when teenage gangs physically took over buildings. You paid the rent to them. If they decided to take over your apartment for a clubhouse, they'd kick you out into the street. They were really badass people—mostly Spanish. Everybody was afraid of them.

We began interviewing the superintendent of the building. He was a nice old Spanish man. However, his wife kept telling him to shut up. I'll never forget, he's sitting at his kitchen table and he says, "What am I to do? This gang . . ."

I said, "You have to tell me who they are and where are they."

"I don't want to get in trouble. They'll beat me up." They were also killing people.

As he's talking to us we hear, crash, bam, crash, crash. I ask, "What's that?"

The super said, "They're breaking windows. They feel like breaking windows."

"What?" I get mad when people do things like that. "That's it. Come on, Al, let's go check the building out."

As we are walking up the stairs we meet five guys coming down. "Hi guys, how are you doing?" I always say hello.

One of them asks, "Who are you?"

I said, "Bombero. We're here to check out a fire you had on the top floor. Where are you guys going?"

"Why do you want to know?"

I said to myself, "Okay, there's five of them, and the more you push teenagers the less you're going to get." Then I said to them, "I'm just curious. Everything okay?"

"Yeah, no problem, man."

"No problem, guys. Have a good day."

•

We walk past them. As we get to the third-floor landing we smell smoke. I said, "Holy shit."

We get up to the fourth floor and find a vacant apartment on fire. It is just in the incipient stage. The flames are about 12 to 15 inches off the floor, in a bunch of mattresses.

I yelled to Al, "They just did this fire." Then I get on the radio and tell the dispatcher to send in the first due units on the box.

"Ten-four fire marshal, box transmitted. Thank you very much."

We turn and run downstairs. On the third floor we again meet the five punks. "Hey, what are you guys doing? I want to talk to you."

"Hey, fuck you, man."

Al and I draw our weapons. "That's it, all of you against the wall. Hey, I said on the wall, mother."

They all looked at the python. Then they all get on the wall. One guy said, "Hey man, I didn't do anything."

"You're all going to jail."

I get back on the radio and I tell the dispatcher, "Fire marshal requesting a ten-eighty five," which means you're not in trouble but you need a backup.

"Ten-four, fire marshal."

Then I said, "Al, keep your gun on them. If they make a move shoot them."

We start to pat them down to let them know we have them under control.

Behind us is the staircase down to the second floor, but it has an intermediate landing where you have to make a U-turn before you get to the next floor. I hear voices on that landing.

"Al, watch them."

I go to the top of the stairs and there is the rest of the gang. Their leader is a girl with a leather jacket. She says, "Hey man, what you doing with my man?"

I said, "Just get out of here. If you want any trouble you got it."

•

"Fuck you."

The gun is still in my hand. I said, "Don't come up here. You come up here you're all going to be arrested."

She says, "We're going to come up there and kick your ass for your gun."

I get back on the radio. "Dispatcher, make that a ten-seventy five," because the fire is getting out of control. I can hear it dropping down and the smoke is starting to build up.

"We're in trouble—we need additional police backup, forthwith."

"Ten-four, fire marshal."

With that the girl starts to come up the stairs. I point the gun at her forehead. She looks at me and says, "Shoot me." These people love to challenge you. They know that you have no right to shoot them.

I said, "Now get out of here."

She starts egging the other guys on. Al is getting nervous, because he has five prisoners and only one two-inch gun.

"Al, if they move, you shoot them right in the head. Don't even go for the body. I don't want them to move after they're shot." You say those things to instill some fear of God into them.

You're always afraid because you know what *you're* going to do, but you never know what the new guy is going to do.

We hear a lot of screaming all of a sudden. I go into a combat crouch and say, "I'm going to blow you away sister, trust me."

She didn't come up the steps. She wasn't too sure of me.

I said, "Come on, trust me. I am not going to lie. I got a fire above me and you below me."

The next thing I know, as I'm watching her, these firemen come up the stairs and push her out of the way. They're not even paying attention. When they arrived they

•

saw the fire coming out the windows, so they have their minds set. They just said, "Oh, you got them good. Fuck them."

In the meantime, I'm saying, "Uh, uh," but I don't want to tell them "Hey, listen, stop the line for a minute, I need help here." What are they going to do? I got a gun and if I can't help myself then too bad.

Then I hear on the radio another fire marshal who comes rolling in. He says, "Hey, Jack where are you? The cops are looking for you."

"I'm on the third-floor landing."

"We're coming."

They come up. Again they were terrific.

Here's the worst part: While the fire is going on, we're in the street putting the five punks into police cars, when all of a sudden the cops go running back into the building. "What's going on?"

"They got a cop in there."

I said, "What, are they crazy?"

The girl and her crew jumped the last cop on the way out.

When we got to the precinct I separate my prisoners and began to question each one individually. I was looking for their leader. He's the one whose body language says, "You piece-of-shit fireman." That's the guy you have to break down. The rest of them are all bullshit.

To make a long story short, I grabbed the leader and said, "I'm gonna tell you something pal, you're going, arson two, second degree, menacing a police officer, threat to my life. But what's sad about it is you have a couple of sixteen-year-old guys there, and you're what—nineteen, twenty years old? It's a shame they're going too, because they're going for the same crime."

"You gotta prove it man."

"I'll prove it. Don't worry about that, but nobody is walking out of here today."

"Well, that's not fair."

•

"What do you mean, that's not fair? I know why it's not fair, because, sucker, you did the fire, and you're going to take four guys with you. You're not a man. If you were a man you'd be the biggest guy in the world if you stepped up and said: I did the fire. Because if you say that I'll cut the other four loose. I don't care. I only want the guy who actually struck the match, but if I take five of you it makes no difference to me. I get paid no matter what."

"Can I talk to them for a minute?"

Two of them were crying. We checked the records. They had never been arrested before.

You have to play all the angles in this job.

He comes back and says, "I didn't mean to do it, man."

I said, "It doesn't matter. You want to say that. Hold it, don't say anymore." I pulled out a pad. "I'm going to put that down, that you didn't mean to do it, because that will be important six months from now."

I said, "You're going to take the whole hit, okay?"

"I'm man enough."

I said, "I thought you were. I like you, man. You want a soda first? We're not in a hurry."

"Okay."

I buy him a soda. It costs me more money for an arrest than I make.

He told me that he set the fire. He's been setting fires because he wants the firemen to come in, chop up the floors and the walls, so he can take the brass pipes out and sell them.

They locked up five more, including the girl, from the police side of it.


I was working in Manhattan when we had a fire in the vestibule of a big apartment building.

We had an eyewitness who told us that the fire was set by a tall, leggy, black woman who wore a big, blond wig.

•

The tenants of the fire building knew the woman. She gave favors to guys in the parking lot across the street. The tenants kept calling the cops and yelling out the windows to drive her away.

One day she got mad and said, "I'll fix your asses." She came back with some gasoline and poured it on the vestibule door. It was a nothing fire, but it was in an occupied building.

We came around every night, for five nights after the fire, looking for her. Finally, this great little old woman called the office and said, "She's here."

We drove over there. The blond spotted our car. She ran for her car, jumped in, and took off. We all go east on Forty-eighth Street. She was barrelling along. We have the lights and siren on. We forced her over at Madison Avenue.

She gets out of her car, and I cuffed her. Then I threw her in the backseat of my car.

A second marshal car pulls up, and I say to them, "Get her car out of here—impound it."

My partner is driving. I am in the backseat with the prisoner. I get on the radio and say, "This is Squad 47 Alpha. We are removing one female to Manhattan central booking."

She laughs at me, and says, "Man, you don't know what the fuck you're doing."

I said, "What?"

She says, "You got to search me a little better."

She was a male transvestite.

I had to get back over the air and say, "Forget that female; it's an *it*."

I got stuck in central booking with her for something like 30 hours, because she had to go to arraignment.

She went into a cell, while I did the paperwork. She got beaten up by a bunch of guys. She said, "They're hurting me, they're hurting me."

I said, "Oh, goddamn it."

•

"Can you get me a soda? Can you get me cigarettes?"

"All right, all right."

She became my friend.



I was working in Brooklyn with Bobby Murrow again. We get a phone call from a black woman named Gloria Green*. Someone threatened to burn her out of her apartment.

We go to her apartment in the Park Slope section. Park Slope is starting to turn around. It's in a four-story building.

We sit down with her. She is 275 pounds of love and fun. She was the nicest person you ever want to be with— funny, hilarious; but she told us that she's scared to death. Some white boy who works for the landlord told her that they'd give her $200 if she'd move out, and they would find another apartment for her. If she and her children don't, then they can't guarantee that something won't happen. It could be a fire. It could be that her children get hurt.

I asked, "Do you know who this person is? Can you identify him?"

"Yeah, yeah."

I said, "Would you be willing to wear a wire and talk to these people?"

"Oh yeah, they're not going to chase me out."

Her husband was saying, "Honey, baby, do you really want to do this?"

"I'm not going to be chased out of here. Who do these boys think they are?"

The white boy is an Irish guy, in his late teens, named Tommy. He used to live in the building but now lives around the corner with his mother. He is peck's bad boy: A real scummer. I mean a real lowlife: A cop fighter and a drinker. He's not only threatening Gloria Green, but everyone else in the building.

•

*  *  *

The landlord wants to get everybody out so he can renovate the building and turn it into a co-op.

The community board got involved, so this case started to get really big.

We go back to Gloria and tell her to make an appointment to meet this guy. "Call him and say, 'Look, you know, I was thinking about your $200 offer, and maybe you can make it a little more.'"

She calls up. He's going to meet her at 6:00 tonight.

I get excited as hell. I drive down to the DA's office, and get a Nagra, a wire, and a transmitter so we can monitor her in case she gets in trouble.

I have the Nagra taped to the small of her back, but I still have to put the Nagra mike on her. I'm standing there with this little mike in my hand, and I have to tape it to her skin. I could fit my whole fist between her cleavage. She is a big woman. I have to move her breasts to tape the mike in position. I said, "Your husband is going to do this."

She said, "Don't worry about him. Anyway, you've seen them before."

We give her some key words to say if she thinks he's going to hurt her.

Stickevers is in one car with me and we're listening to the whole thing. We have another squad for backup in case anything goes wrong.

She meets him on the library steps on Sixth Avenue and Ninth Street. They have a 45-minute conversation, though he didn't admit anything on the tape, but we know that she's for real.

We get Tommy one more time on tape, but we still don't think we have much.

In the meantime, there are a couple of small fires in the building.

About two weeks later we go to a DA with what we

•

have. The DA listens to the tape and says, "You have enough to grab Tommy, but he never mentions the name of the landlord. I'm going to authorize you guys to take Tommy."

Before we call on Tommy, I go and talk to the landlord in his home just to let him know that we're investigating something.

His home is a brownstone in Park Slope, where all the attorneys live. It has a spiral staircase in the middle of a marble floor. His wife comes out—she is a beautiful woman. He was a good-looking guy. I mean, these are class people with big bucks. They gave me a lemonade with a cherry in it.

He had a fire in the building where Gloria lived. It could have been related, but it was months before he tried to get her to move out.

He said, "I thought you guys already investigated it."

I said, "We're investigating now. Do you know any reason why somebody would want to burn your building? How much insurance do you have?"

"Do I have to tell you these things?"

I said, "You can tell me here or you can tell me down at my place. I don't care. I'll subpoena you."

"Oh, you guys have subpoena powers?"

I said, "Better than the police." I let him know that we weren't playing around, although I really felt as if I was playing Columbo and enjoying it.

A week later we go after Tommy. We get psyched. Six guys go to his apartment, because he likes to fight. We are wearing flack vests. We knock on the door. His mother yells, "Yeah, who is it?"

"Police."

"Well, what do you want?"

"Does your son have a 1967 Chevy?"

"Yes."

"Well, we have to talk to him. We think his car was involved in an accident."

•

She opens the door and looks at us.

We're armed to the teeth, but we are not showing anything. I'm holding a gun behind my leg. I always do that whenever I go to a door where I might be shot. I'm going to go down shooting.

"Can we talk to him?"

"Well, he's not home."

We're all looking over her shoulder. She is a small, old woman.

She says, "Hey, you want to come in and look around?"

"You don't mind?"

"No."

We rushed in the place. Then she sees the two guys with shotguns. Tommy is not there.

The precinct knows that fire marshals are going to the house to talk to somebody, just in case there is gunfire.

Tommy goes to the precinct and says, "I understand you guys have been looking for me. What did I do wrong? I didn't do anything."

The desk sergeant said, "I don't know anything about it, but Fire Marshal Carney is doing an investigation."

It is 1:00 in the morning when I get a phone call at the base from Tommy.

"What do you want?"

I said, "I want to talk to you. Where are you?"

"I ain't telling you. What, do you think I'm stupid?"

"Tommy, whether you talk to me tonight, or you talk to me tomorrow morning, or you talk to me next Wednesday, it doesn't matter to me. Good-bye."

"No, what do you want me for?"

"I'll tell you when I see you; but I'm going to tell you this much, you're going down hard unless you volunteer. Guaranteed."

"What am I going down for?"

"You're going down, I don't give a shit. You're pissing me off."

•

"But I didn't do anything."

"Nobody is saying you did anything, but I'm not so sure that you understand the severity of what we're doing here."

He said, "Yeah, what—shotguns and vests. What did you think, I was going to shoot you guys?"

I said, "We're hoping for it. That will end our problem."

"Man, I think you got the wrong guy."

"Well, then come in and talk to us." I made it as if he committed the biggest bank robbery in the world.

"All right, when do you want to talk to me?"

"Tomorrow morning, nine o'clock at the precinct. This way we go into a room, we sit down, and I hear your story."

"Uh, what if I don't show up?"

"Then I worked a couple of extra hours for nothing, and I'll catch you tomorrow night. I know you got a girl in Sheepshead Bay." You always let them know that you have information about them.

I hung out in a bar in his neighborhood one night. While I was there I asked, "Anybody know where Tom is? I've been looking for him for three days. He told me to see him. I have a job for him."

One of the patrons said, "He's probably out in Sheepshead Bay with his girl."

We get him in the precinct and talk to him. I told him, "We got you: Extortion, threatening, solicitation, all kinds of shit."

"What are you talking about? I don't know what you're talking about."

I said, "Hey, who's the man who wants to pay these people to get out of the building?"

"I don't know what you're talking about."

"Okay. I'm going to arrest you."

"Don't do that to me. I'll tell you. I didn't want nothing to do with this."

•

He gives me the landlord.

There is no real arson here, but there are many other felony crimes.

Tommy gave it up in order not to go back to jail, because he is on parole for assault and robbery. He has one year left on his parole, but if he doesn't make it he has to go back and serve the six or seven years remaining on his prison sentence.

I told him, "We can go easy on you, but you have to wear a wire."

"Okay."

We're in a car near the southwest side of Prospect Park. We have three squads close by in case something goes down.

Tommy's wearing a wire, and we are monitoring him. He goes and knocks on the door of the landlord's house.

We hear the door open. The landlord says, "What the fuck are you doing here?"

"Hey man, I got to talk to you."

"You're not coming in here. Don't come in my house. What are you doing here?"

"Man, those fucking fire marshals are looking for me. They came to my house with bulletproof vests."

"Who was it, that little fuck Carney with his black partner Murrow? That tenacious little bastard. He doesn't let go, watch him. . . . Get the fuck away from me."

"Hey, what am I going to do? Give me some money so I can get out of town for awhile, until things cool down."

I'm as excited as hell.

All of a sudden the tone of the landlord changes, "Hey, you wearing a wire? You wearing a tape?"

Tommy says, "What do you got there? Put that gun away man. Put the gun away."

It took three seconds for us to say, "Squads roll in."

Then we hear the landlord say, "Get away from me," followed by the sound of a door slamming.

Then we hear heavy breathing, followed by Tommy say-

•

ing, "I'll meet you at the corner. He had a gun. He had a gun."

I call the squads off right away, because I don't want the landlord to know that Tommy really did have a wire on.

We bring the tapes back to the DA. I'll never forget the day. It was a bright, sunny morning around 10:00 or 11:00, when the DA gives us the green signal to lock up this bastard landlord.

Tony Romero, who is the chief in Brooklyn, says, "Jack, nice job."

I said, "Well, this is a cowardly bastard who hires punks to get poor people out. I want this guy."

"You got it, but we're going to send two guys with you, John Knox and Jim Callender.

We roll in an unmarked car, all dressed up like gentlemen. We knock on the door. The maid opens it and says, "Yes?"

I said, "Is Mr. Murphy* home?"

"Oh yes."

His wife comes to the door, "Hello, Fire Marshal Carney," with a distinct, beautiful nightingale voice, "how are you?"

"Hiya, ma'am. How you doing? This is Fire Marshal Knox, and this is Fire Marshal Callender." We walk into the vestibule. "Can I speak to your husband, please?"

"Why, he's busy right now. Maybe I can help you."

"No, we have talk to him." We had a warrant in case we needed it, but we didn't want to use it.

He then comes out, and I say, "How are you doing, sir? Your name is Mr. John Murphy?"

"Yes, that is my name. You know my name."

"Good. You're under arrest."

"Don't be absurd."

I'll never forget him saying that.

"You're under arrest. Put your hands on the wall."

He looked at his wife, and said, "Jane, call my lawyer."

•

Turning his head back to me he stated "I don't put my hands on the wall."

John Knox pushes me to the side and yells, "Hey shithead, you're under arrest," and cuffs him before I know it.

Murphy begins to say something when Knox cuts him off, "You already talked to John, right? Now shut up. I don't want to hear from you. You're under arrest, scumbag."

His wife starts crying hysterically, "What are you doing to my husband?"

"He's locked up, lady."

"Where are you taking him?"

"Call your attorney."

Out we go. We get in the car. I am in the backseat with my prisoner. We open up the windows, and start the car. They hit the siren—four or five long ones. People start looking. We put the red light on, and drive out at exactly two miles per hour.

I was never called to court. He probably copped a plea for a misdemeanor.

Later a detective called the bureau and said, "You got a fire marshal named Jack Carney? Well, the word in the street is that there's a mark out on him, and it is coming from someplace in Park Slope." However, we couldn't link it to our landlord.

My partner was on vacation, and I was working as John Stickevers's aide this particular night. We're sitting in the kitchen of the Queens base in Fort Totten listening to the fire radio. We hear a telephone alarm for a box on Gladwyn Avenue, which is in Flushing, Queens.

John Keittleberger is also in the kitchen. We're having coffee. We hear the engine give a 10-75. Well, Keittleberger says to me, "That is a nice neighborhood."

I said, "Well, let's take it in. They're probably going to

•

call us later on anyway." Although fires in nicer neighborhoods are usually accidental.

We went in with two cars. As we rolled in, the engines were hooked up to the hydrants and stretching lines. We got there very quickly.

I parked the car down the block and walked to where there was a big crowd. It was about 11:15 on a weekend night.

The chief comes over and asks, "Are you guys marshals?"

"Yeah."

He says, "The building is vacant. They sold it, and they're waiting for the people to move into it, but they do have security."

Keittleberger says, "Jack, tell the dispatcher that we're on the scene, and we're going to be investigating this fire. We'll make it a 10-41, code one.

A security guard shows up for a midnight shift. He is from Jamaica, a nice guy. He says, "I don't know where the other security guard is."

I never go to the fire building when the fire is going on. I disappear into the crowd. I don't display my badge. I just mingle. I'm a creature of habit, I really am.

I'm looking around, when Keittleberger says, "Jack, what are you doing?"

"I'm just hanging out." He laughed. Then I notice that people are laughing and drinking. "There is something wrong here. Besides the building being vacant, the neighborhood is too happy."

John Stickevers said, "Jack, let's stick around. We'll help them with the physical and get some interviews."

I bring the car up opposite the fire building. The fire is now out, and the engines and ladder trucks are taking up.

Somebody in the crowd yells, "This wouldn't have happened if they obeyed the court order."

Well, that's like a shot of adrenaline to me. I think we got something.

•

I have the trunk open. I'm taking my shoes off and putting my fire boots on, because I'm going into the building. A woman comes over, and I ask her, "What court order are you talking about?"

She raises a piece of paper, and says, "They're not supposed to be in that building. If they weren't in that building there wouldn't be a fire. The security probably had an accident and set the fire."

I notice that every time I look at this other young woman, she would look away. I said to myself, "She knows something. That's number one."

Then I see that the woman with the paper in her hand is doing a lot of talking, and she apears to be very happy. That's two.

I go into the building, and grab my boss. I said, "John, we have something going on."

He says, "Yeah. Well, I just found out that this house has been sold to the City, and they are going to transform it into a home for boarder babies. They are black babies whose crack-addicted mothers have abandoned them."

I said, "Oh, terrific. This is going to be big."

John says, "Tell me about it."

I leave the house and start walking up and down the block. I catch the eye of a heavyset man sitting on a stoop.

In the meantime, the woman who earlier turned her head away from me four or five times goes inside, but I know what building she went into, so I know where she is. I also knew where the loudmouth went.

I go back to the fellow on the stoop, and I yell to him, "Excuse me, sir, can I have a glass of water? I'm dying of thirst."

He says, "Yeah." So, I go up to the stoop. He sends his mother in the house for the water. He is not looking at me. He's tapping his feet. Finally he says, "Vigilantes. I've never seen anything like it in my life. . . . Don't talk to me."

•

"Okay, no problem pal, no problem. I've been there before. Have you ever seen the movie *Rio Bravo*?"

"Exactly. John Wayne."

I said, "Good."

His mother returns, "Oh, thank you for the water, ma'am."

I then asked him, "How am I going to talk to you?"

"I'm leaving for Baltimore." He is an attorney in Baltimore, Maryland. He came up to visit his mother for the holiday.

"Give me your phone number." He gives me a 604 area code and the number, but I don't want to take a pad out and write it down. So I said, "Repeat that." He does, and I quickly walk down the block and into the fire building. Then I write the number down. I say to myself, "This is good, we're getting into this one."

John comes back to me and asks, "You got something going?"

"I'm working on something."

"It figures you would."

Our physical investigation showed that a flammable liquid was poured on the first floor and burned down into the basement. We finish at the fire scene. It's now 4:00 in the morning and everybody's gone to bed. They had their fun. We go back to the base and type up what we have.

The next day, the mayor and the world showed up on Gladwyn Avenue. It gets into the newspapers.

I get a call, Jack Keittleberger. He says, "The police are taking over this case."

"All right, let them take it over. I have something I want to check up on anyway, since it's still a fire marshals' investigation."

They called me the next day to come on into work and start working on this. In the meantime, I spoke with my attorney friend from Maryland.

He said, "I just want to tell you that the people in the

•

neighborhood didn't want black people in there, but I don't want to say anymore. I don't want my mother involved. If you protect my mother, and the City of New York buys her house, then I'll testify."

I said, "Okay. Good." Then I said to myself, "Oh man, I love this."

In the meantime, A and E, the police department's Arson and Explosion Unit, is taking over the case. But there was so much heat over this case that a joint task force of fire marshals and police was formed.

ADA Charlie Testrogrossa, the Bureau Chief of Arson and Economic Crimes in the Queens DA's office, is assigned to head up the investigations.

A group of us go down to Maryland to interview my witness. We go to his law firm, but the law firm says that their employee is now their client, and we have to talk to them before we talk to him. I said, "Wait a minute. This isn't a court of law; we're trying to find somebody. You're hindering a prosecution." It was a little bit touch and go there. What turned out was that we had to call the fire and the police commissioners and the mayor of the City of New York, to see if we could strike a deal and buy the house from the mother and give her police protection. Well, the City wouldn't go for it. Police protection was no problem, but they're not in the real estate business. So, he wouldn't talk.

Later, we went over to his home, without the presence of his attorneys because he was an attorney, and asked him one more time to help us.

He said, "I'm not supposed to be talking to you," but he gave us a description of who was seen walking away from the building just minutes before the fire. He wasn't friendly with the neighbors, since he lived in Maryland, so he didn't know their names. But he knew that they were involved in getting the security guard out of the building by breaking windows in the house earlier in the night.

•

We came back to Queens, and we were going to do a massive canvass; that was a smoke screen. I interview all the suspects over a period of two days. I didn't have to read them their rights, because they weren't being arrested.

Lo and behold, one of the suspects in the case is the same woman who wouldn't look at me. This detective and I go into her home, which is directly across the street from the fire building.

Both she and her husband are executive officers for a bank. They are educated people.

We sit down at a nice table. She made a fresh pot of coffee, but everything she did was jerky. We start to talk, "What did you see that night? What time did you go to bed? What was on television?"

You try to bring people back some hours before the event. "What did you do the morning before the fire? What did you do that afternoon?" Now they start thinking about what they really did, so when they come to their story-telling it gets jammed, because the real things are there.

I said, "Listen, I just want to tell you something. I don't care what you tell me, it's okay."

Now she is looking at me, and she's shaking. She is sitting directly across from me. Her husband is to my left at this big rectangular table, and the police officer is to my right.

"I don't care. I'm writing it down. But I'll tell you this much, if I find some of the things here are not right I'll come back, and you're going to be in trouble, no question about it. Because there is no more talk after this. This is a friendly interview."

The husband puts his head down, and he says, "I can't do this."

I say to myself, "Shit, I got him already." She breaks down and cries. The detective almost falls off the chair.

The husband went on to tell us the whole story of how the others planned it, and how they wanted him in on it,

•

but he really didn't want to do anything, though he ended up being a lookout.

I didn't want to make a collar and hustle him out of the house. Instead I reported back to my bosses and the ADA.

Three hours later, our suspect left his house with his wife, supposedly to go shopping, but in reality they drove to our base. The police and the ADA were there. We sat down with him again and took a second in-depth statement. He was an excellent witness. We asked him if he was willing to wear a wire.

He said yes. Later he called the other suspects up. He got three people on tape saying, "They have nothing on us. They're just fishing."

We went to the grand jury. The jury handed down five indictments.

During my investigations I went to all their homes, and their homes were no different than mine. Yet they were going to be arrested on felony charges and go to jail. They worked for years for these homes, but they took the law into their own hands—classic vigilantes. They felt the value of their houses would go down, and they panicked. They were right about that, because when our witness wanted to sell his house, he almost had to give it away. Were they right in what they did? No, they definitely were not right.

I really felt bad arresting them, because they had never been through it before. One of them broke down and cried. I told her, "Look, I'm not here to hurt you. I'm not here to make you embarrassed. Just do what I want. I have to put cuffs on you." That is probably the most terrifying thing to anyone—when you make an arrest and put cuffs on them. I don't care who you are. Once you take cuffs out they know that this is for real.

Before we even went to trial, the house was refurbished—there wasn't much damage. The babies are in it, and they have police protection 24 hours a day. The City

•

was talking about more boarder baby homes around the Queens area, but they haven't opened any.

The neighborhood was there supporting the defendants. The neighbors felt that they were martyrs. They were doing something good for the neighborhood. I believe there was money collected for them for their defense.

I testified in court that all fires start with a small fire. They all have to start with a match or some kind of ignition. I said, "I think what people are forgetting in this courtroom is the danger in the response of speeding fire trucks, and the danger for firemen stretching lines and crawling into that basement. Not to mention the danger to the houses on either side." I testified for days, because most of it was my investigation. I shouldn't take all the credit, because there was much good work here.

All five were convicted and sentenced to serve up to three years in the state prison.

That was my last case before I retired a little over two years ago. I'm very fortunate to say that I had a terrific career. As a matter of fact, I think I made a mistake retiring. If I could, I would go back tomorrow. My wife wouldn't like it, nor would my sons, because they figure that if anybody's going to get shot in our family it's going to be me.

•

# 4

•

## SUPERVISING

## FIRE MARSHAL

## ARTHUR MASSETT

I became a fire marshal on August 20, 1977. I guess I was getting bored with the firehouse. The only time you got out was to go on an alarm, and after awhile all the fires just seemed to look the same to me. I wanted something new, that's all. The fact that I had been a police officer prior to becoming a fire fighter also had an influence on my decision, because I enjoyed being in the street on my own.

When I first came into the marshal's office it was right after they burned Bushwick to the ground. Mayor Beame said we had to have more marshals. So, they put 50 new marshals on the job.

My first real job as a marshal was for a fire in a public school in the Bronx, on 153rd Street. The fire was going on during the World Series. The

•

Goodyear blimp was flying overhead, because the fire was right up the road from Yankee Stadium. It made national television—the South Bronx burning.

It was one of the largest fires I have ever seen. Howard Cosell put it "under control" at the third alarm, but it went on for two more. It was burning so long that we took a break and went over to a nearby firehouse. That's where I saw it on the TV.

We investigated that fire for maybe two weeks. We had an idea right from the beginning who did it. We had three girls: One Irish and two Italians who were still living in the projects down there, and they saw it happen. We met the girls when canvassing the crowd of people looking at the fire. My partner, Jack Bowens, who is now in our juvenile unit, and I would just wander up and down the sidewalks, talking to people, "Do you know anything about this? Did you see anything?"

The girls said, "Yeah, we saw these teenage boys go in there, and we saw them run out. They brought in these square (five gallon) cans, but they ran out without them, and then this tremendous fire happened."

In those days, they were using a combination of gasoline and either diesel fuel or fuel oil from an oil burner, because the latter two tone down the explosiveness of the gasoline.

The girls rode around with us for a few days trying to spot these guys in the street. The teenagers had the typical description of the day: sneakers, blue jeans, and T-shirts. But we never found the three kids who started it.

My first arrest was a guy named Willie. I still remember him. Willie was living in a vacant apartment on the top floor of an occupied building on Eighth Avenue and 138th in Harlem. It was in December 1977, and it was cold. He was burning wood inside a metal kitchen cabinet that he had on the floor. The heat conducted through the cabinet, into the wood floor and started the place on fire.

The fire department came, put it out, and told him not

•

to do this anymore. The people in the building threw him out.

Now Willie was going to teach them a lesson. He came back, went into his apartment, and set the place on fire. But it was a much bigger fire than the first one. And since it was in a top-floor apartment, and the fire department comes in and puts the fire out with water, all the bottom floors had water damage. Within two weeks it became a vacant building.

We catch Willie. In those days we used to stand arraignment with the prisoner in Criminal Court on Centre Street in Manhattan. You sat down there for hours and hours waiting for your case to be heard. The sergeant would call us up in groups of 10. He would say, "Okay, go on upstairs and get your prisoner." Then you would bring the prisoner to be photographed before putting him in the holding pen behind the courtroom. You'd go out into the courtroom and wait for the bailiff to tell you to go back in for your prisoner. You had a little bakery number ticket which you presented to the guard, and they'd bring out your prisoner.

Now, I'm here to save the world from arson. We get Willie, and we bring him before the judge. It's rapid-fire arraignment. They plea-bargained the case right in front of me. I had locked Willie up for arson two. They plea-bargained down to attempted arson four. Arson four is the lowest of the arson statutes. It is a class D felony. It has to do with reckless damage to a building. Let's say you start a fire in a charcoal grill, but you happen to start it inside your living room. You use charcoal fluid, it catches onto the drapes, takes out the apartment and the rest of the building, and people lose their belongings. Well, you intentionally start a fire, though your intent was not to damage the building, but because of your recklessness you did in fact damage the building. That's arson four. Now, you can never *attempt* recklessness according to criminal law. You either are reckless or you're not, but for plea-

•

bargaining reasons they let him plead to an attempt at recklessness.

He walked out the door with six months' probation. And I'm standing there saying, "goddamn it." I was looking to put this guy away forever. Only in New York.

I walk out of the courtroom, and Willie is standing in the vestibule waiting to leave the building. He comes over to me and says, "Excuse me, officer."

I said, "Yeah?"

"Can I borrow a dollar? I got no money. I want to get back uptown."

I just shook my head and said, "I'm sorry, I don't have any change." Then I walked away.

I don't know what ever happened to Willie. I hope he kept his nose clean. I'm willing to bet that he didn't. But that was my first arrest.

In this business you come across many unexpected things. I remember being called to a fire in a chicken barbecue on 157th Street and Broadway. We get into the restaurant after the fire was knocked down. We find what looks like burglar's tools lying in the back room. There's an electric drill, chisels, and a hammer. The night manager is also there. He is an 18-year-old kid. He's being very helpful to us. His uncle owns the place, but according to the company's rules you are not allowed to hire relatives.

We look at the safe that is also in the back room and we see marks on it. Now we don't know if we have an arson fire to cover up an attempted burglary.

The kid, who's being very helpful, claims he locked the store up and left, but there is just something about his story that doesn't make sense. So we started asking questions around the neighborhood.

There is a hardware store across the street from the restaurant that is still open. We go in to talk to the owner. He says, "Oh, you're here about that fire over there. I heard about it. The kid that works there was just in here last week. He bought a whole load of stuff."

•

"What did he buy?"

"He bought an electric drill, some chisels, and a hammer." All the things that we found lying in the back room.

We go back to the kid and present him with the facts. He said he left at a certain time, but at the subway station the token booth attendant, who knows the kid because he travels that way all the time, said he didn't come that way that night. We also told the kid that we knew about his recent purchases.

Eventually he confessed.

I asked him, "Why did you make the fire?"

"Well, I made the fire to cover up the burglary."

"What's with the burglary?"

"I had to get into the safe, because I needed money."

"What did you need the money for?"

"For my girlfriend's abortion."

"Your girlfriend is pregnant?"

"Yeah, I got my girlfriend pregnant, and she needed money for an abortion, but I didn't have any money. So, I had to steal the $300 it would cost."

"Gee, that's really too bad."

We go down to court, and as we are going through the arraignment process who shows up? His mother and his uncle, who by the way lost the store because he hired his nephew. They're standing by him, when in walks this young girl, his girlfriend. To make a long story short, thank God he wasn't a bad kid. He got a year's probation. But the girlfriend was never pregnant. She told him she was pregnant, because they were starting to break up, and she thought that she was going to lose him. So, he goes and torches his uncle's store, screws up his life, screws up his uncle's life, and it was all because of a lie.

I never understood the crime of arson. It's a dirty crime. It is such a spectacular crime. The fire itself alerts everyone in the world—well, not everybody in the world. Here in the Bronx some people could not care less about fires. Yesterday I went to a second alarm over on Walton Ave-

·

nue. I was going out to get something to eat when I said, "Ah, let me take a ride over there and see what's going on." It was on a block off of 170th Street, which is a pretty busy intersection—a shopping area. Here is this apartment house with two or three floors burning. The flames coming out the windows. There is smoke all over the place, and people continue to shop. Right down the street is the public school. The school yard is packed with kids playing stickball. There is a big fire going on, but nobody's paying attention to it. Life goes on.

Working in the street is both easy and difficult for me. And the reason I say that is, just look at me with my Irish face and my red hair.

One time we were to rendezvous with my first supervisor, Paul Drabeck. He told us to meet him at 170th Street and the Grand Concourse. It was a nice night, so when we got there I got out of the car. I was leaning against a mailbox when he pulled up. He was laughing. I asked, "What are you laughing at, Paul?"

"You know, you fit into this neighborhod like a palm tree." In that respect it was hard, because I could never ever blend in with the South Bronx.

Other than that it was easy. You never had to explain to the people who you were. You got out of the car, walked up the steps, and people just got out of your way and left you alone. They didn't know what you were or who you were, but they knew you were "The Man." Whether or not you were the precinct detective, or the DA investigator, or the fire marshal, they knew you were somebody who was there to lock somebody up, and they didn't give you any grief. They didn't confuse me with the social worker or anything like that.

I enjoyed the street. Being inside, I miss it. Sometimes I'll just put the pencil down and get out to a few fires; really buffing it.

•

What I miss most about the street is the excitement. You never know what's going to be happening around the corner.

I think it's comical some of the stories you hear from people out there. I remember one night, I was working with Neil Lagatta, and it was our last tour. I was going on vacation, and he was flying up to someplace in Massachusetts to meet his wife. We both had someplace to go, but we get this fire in Harlem. A mattress fire.

We go up to the place, and we're looking at the building. It's half occupied. I said, "This is going to be a ground ball. We're going to be in and out of this thing in nothing flat."

We walk in. We thought the fire was in a vacant apartment the way it came across on the radio, but there's this woman standing there. So I said, "Could you tell us where the fire was?"

"Yeah," she said, "the fire was in my apartment."

I said, "Oh boy. We got an occupied apartment. Okay, give me your name."

And she says, "This is a jive-ass fire. This is really a jive-ass fire."

"Well what's the matter?"

"That goddamn boyfriend of mine, he did it. I know he did it."

So I said, "It's three o'clock in the morning. Were you home?"

"No, no, I was down at the club, but I know that guy did the fire."

We don't have an eyewitness, so we're not about to worry too much about this one. But I said, "Let's go up and look at the fire." So, we all go up, and the mattress is burned. I asked, "How do you know it was your boyfriend?"

"Look at that broken window." Still no legal evidence to lock this guy up.

I said, "What do you mean the broken window?"

•

"That jive-ass motherfucka, he bust out that window, come in, and set my house on fire."

"How do you know that?"

"Well, he stole my goddamn radio."

"What kind of radio?"

She describes a boombox.

"Well, we will make the report out, and we'll go talk to your boyfriend. Do you know where your boyfriend lives?"

"I don't know where he lives. He's out in the street all the time."

We're figuring this is a grounder again, we're done with it—in and out. We are getting ready to leave, but this woman had ears like you couldn't believe. She could hear the faintest sounds. She says, "What's that?"

My partner and I look at each other and say, "What?"

"Did you hear that noise?"

I said, "No, I don't hear a thing."

"I'll bet you that's him." Now she walks out of the apartment, into the hallway, and yells down the stairs, "Who's that?"

We hear this voice coming from downstairs, "It's me, baby."

Now I look at my partner and he looks at me and we're saying, "Oh God, what have we got here?" But we still have no connection between the fire and this guy.

"It's me, baby."

"Why did you burn my motherfuckin' apartment?"

"Baby, I'm sorry." Sort of a spontaneous admission. "Baby, I did it 'cause I love you."

"You love me? Why'd you steal my fuckin' radio?"

"Baby, I'm bringing it back."

Now we got the evidence. We know this guy was in the apartment. He comes up the stairs, and he walks right into our arms. He's got a cut on his arm from the broken window glass, and he's got the radio in his hand. The guy fesses right up.

•

Now Neil and I are looking at each other. I'm going on vacation. He's going to Massachusetts. It's 3:00 in the morning. We're in Manhattan, and we're going to be with this guy forever. So here are two fire marshals standing in the hallway, with this guy cuffed, flipping a coin. The loser has got to take the collar. We wanted no part of it. That is the bad thing about this job. It can screw up your social life, because you just don't know when you're going to get home.

I had a fire on the East Side of Manhattan in a good neighborhood. The fire came in at 2:00 in the morning and I didn't get home until 11:00 the next night. It was in a new-law tenement: A rent-controlled building that had been taken over. One by one, as a tenant would move out, the landlord would come in, renovate the entire apartment, and co-op it. But there were a few old-timers left in the building. One of them was a couple with a 20-year-old retarded daughter.

Their apartment was like the Collier's mansion: The place was just loaded with books and newspapers. They had a tremendous seven-room apartment and every room from the floor to the ceiling was jammed with books, magazines, and newspapers. They had this little vestibule, right outside their apartment, where they kept additional stuff. The neighbors would complain that it looks like garbage. The landlord kept saying to them, "Get this stuff out of here."

One night, somebody sets the vestibule on fire. The fire gets through the door, and it takes off in the apartment. The occupants of that apartment got out via the fire escape. But the man upstairs didn't get out, because he fell asleep with a bottle. He was an alcoholic. He dies in the fire.

Now, because of the location, a well-to-do neighborhood, Lexington Avenue and Ninety-Second Street, all the

●

stops were pulled out. In comes the 19th precinct. The place is crawling with detectives.

We worked on just getting the physical. The sanitation department had to bring in dumpsters, because we were throwing the stuff out into the street. We filled up the intersection and they had to bring in a pay-loader to scoop the stuff up.

This guy claimed he was a book collector. I don't think a collector collects them in his apartment to the extent that this guy did. He had bookshelves up the walls to a height of eight feet, and then he would build a shelf across to the other wall and he would stuff books up onto that shelf. You were walking through a tunnel: You were between and under books, magazines, and newspapers.

Can you imagine the amount of fire once this thing got cooking? It was blowing out every window in creation. There were a lot of victims with smoke inhalations, but there was only the one death.

It took us hours and hours and hours to dig this place out to make sure what we had was in fact the origin and that it didn't start in the apartment and burn out to the hall. We had to show a clean area below what we determined was the point of origin.

Then of course we had one witness: This young girl who was visiting New York and living in her aunt's apartment. She was out bouncing. She came home at 1:00 in the morning, with half a package on, and sees this stranger loitering outside the building. She goes into the building, and a half hour later we have this fire that burns out the third floor. It puts a monkey wrench into the whole thing. Here we're thinking maybe it's a tenant, when out of left field comes this girl saying she saw some guy standing in front of the building.

Some bartender around the corner said he saw a suspicious man. Why would a so-called stranger go into a locked bulding, go up to the third floor, and set this pile

•

of rubbish on fire? Anyway, we canvassed everyone in the building. They all had alibis. Most of them were trapped by the fire themselves, so you eliminate them because usually people don't set fires to trap themselves. The unknown stranger became more and more of a suspect as we went along. But the unknown stranger in New York City could be anybody. We got nowhere with it. The cops finally dropped it, and they were in charge of the investigation because it was a homicide. Under the police fire guidelines that came out of the early days of the Arson and Explosive Squad, the cops were doing the investigation. We were only going in and making the physical examination to determine if it was incendiary or accidental.

At that time, the fire marshals were only allowed to make arrests if the suspect was within arm's reach. What that meant was, if you didn't have to make any extraordinary efforts to search for this guy you would be allowed to make the arrest. But if it was John Smith who lived somewhere over on 149th Street, well, that information had to be turned over to the PD, and they would do the subsequent investigation and get the collar. So for a period of time we weren't doing any arresting. But I don't know if the cops should be doing it. They have enough problems with murder, rape, and robbery. That big blue machine is very politically aware, and they like to keep themselves in the limelight. They want to have their picture on page one. What else can it be? I mean, why else would you want to be getting involved in something like arson?

I had a fire in Queens where three kids were killed. It made the newspapers and it made television. The reporters were there.

The detectives came from nightwatch. The fire took place about 11:00 at night. We were there all night and the detectives show up around 5:00 in the morning in their Burberrys, their Gucci shoes, and their pinky rings. The fire was in the basement. There was about a foot of water down there. We came up the stairs, and we're walking

•

down the driveway toward the street. There's myself and Neil Lagotta with these two detectives.

The television reporter—Melba Tolliver—is there with her cameras. A uniformed cop at the front of the driveway is keeping people out. I hear her say to this uniformed patrolman, "What happened? How did the fire start?"

He says, "I don't know. Why don't you ask the fire marshals? Here they come now."

So, she looks down the alleyway and sees the four of us, and she says, "Which ones are the fire marshals?"

I'll always remember this cop looking at us and saying, "How long have you been a reporter in this town? Don't you know the fire marshals are always the guys with the dirty shoes?" Arson is a dirty crime.

That was a fire where I was pretty sure that it was an older brother who killed the kids. It was out in Queens, nice middle class neighborhood, a frame house, and there were three kids, five, six, and seven years old, dead inside it. It was a bitter, bitter cold night.

When we got there we found out that the oldest kid in the family, three months earlier, discovered a fire in the house they were previously living in, which is three houses down the street. That one was an electrical fire. He discovered the fire in a wall. He alerts mom and dad, and they get all the kids out and call the fire department. Because of this he was a hero. Daddy buys him a new 10-speed bike, and the family moves down to this new house. Now the kid is pestering his father for a stereo.

The father says, "No, you just got a bike. I'm not going to buy you a stereo."

Mom and dad are out of the house; it is Friday night. The kid goes down to the basement and starts a fire on a workbench. His idea was that he would go upstairs, get the garden hose, and put the fire out. It is the middle of January, and the hose is frozen solid. The fire gets away from him. He and his oldest sister get out, but the three other kids die upstairs.

•

This is all in theory, because we never got a chance to really get at this kid. We got this story piecemeal out of his sister. He was 13 at the time, and the youthful offender law wasn't in existence then, so we had to get mom and dad's permission to talk to him. The parents flat out isolated him. But they knew, because we told them what we thought happened. They just sort of looked at us and responded with, "Umm hmm, umm hmm, ah ha, yeah, I see, I understand. But we'd rather you not talk to him." They showed no emotion whatsoever, but they had three kids lying in the morgue.

The kids were roasted. I mean these were not smoke inhalation kids. The fire was in a wood frame dwelling, and it was a complete burnout. The kids were found during the secondary search lying where the floorboards burned through in what used to be the bedroom. Oh, they were just charred beyond recognition—you didn't know if they were boys or girls. You only knew they were children because of the length of the bodies.

I'll tell you something, I have seen a lot of dead bodies since I've been in the fire department, but I just cannot get used to seeing dead kids. When you look at a five year old you know he didn't have a shot, and it wasn't his fault and even if it was, a five-year-old kid doesn't know any better, as we're finding out with our juvenile unit. I can go right up to a dead adult at a fire and look in their nostrils to see if there's any soot in them, and it doesn't bother me. But it bothers me to have to look at a young kid. At that time, my kids were about the same age as these kids, and I said to myself, "This could be me."

I would say that one was probably my most tragic case. I still think it was the older brother who did it. He is now maybe 20 years old, and he knows what he did then. He's been carrying that around. I don't know what's going to happen to him. I'm sure that the parents knew about it whether or not they chose to deny it and not do anything

•

about it. But I hope they went and did something about it on their own.

That fire came right on the heels of another fire where we lost two kids in a South Jamaica home. It was a completely accidental fire. The electric motor on a home freezer overheated, and the fire got into the adjoining wall. The fire was at 5:30 in the morning. Everyone was asleep. Everyone got out except these two kids, and they died. Terrible tragedy, an accident of fate. An electric machine went out of order.

I haven't been unfortunate enough as a fire marshal to have caught that many multiple deaths in fires. The three kids were the most. As a fireman I've been to fires with 12 dead. They were the midtown Manhattan spectaculars. We had many dead bodies in those days. But as a marshal most of my DOA experience has been a lot of single ones.

I had a couple of homicides where the dead body was lying in the middle of a vacant building with a cord wrapped around its neck, and the place was on fire.

Neil Lagotta and I were partners for about two and a half years, and this one job we had comes to mind. It was over in the West Bronx. The neighborhood was totally Jewish, but it changed. Co-op City opened up and a lot of that population moved up to it. The old neighborhood started to get run down. But this old Jewish lady was one of those people that just wanted to stay. She had lived there all her life.

One day someone got into her apartment and smashed her in the head with a ballpeen hammer, threw her on the mattress, and set the mattress on fire. Then they stole her television set. That in a nutshell was the whole crime.

What they didn't count on was that this little old lady was a tough nut, and she crawled off the bed. Her whole head was smashed. One side of it was caved in. But she crawled off the mattress and out to the hallway, where she collapsed and died. But she got out of the fire.

•

As soon as the fire department shows up they see this body with the head smashed in, so they know they have a homicide. We were riding in the Bronx and were practically right around the corner when they called us. The cops eventually showed up because this is a homicide.

There's blood on the apartment floor. Neil also sees blood in the hallway, just a droplet of blood, but not where the old woman was. So, we now know that who ever it was came out of the apartment through the door and didn't go down the fire escape.

Directly across from us are these two Hispanic teenagers. They're out in the hall looking at what is going on like the rest of the curious neighbors. Neil happens to notice on the T-shirt of one of the kids what looks like blood. Then Neil looks down at his sneakers, and the kid's got blood on them. At first Neil doesn't say anything. But he goes over and tells one of the detectives, "Not for nothing, but that kid's got blood on his T-shirt and sneakers."

Before we left the floor they got a confession out of the kid. The kid showed us where he put the TV.

He knew this woman. He used to pick up the woman's groceries. She always gave him, like little old ladies do, two or three dollars, which was a lot of money to this kid, for doing absolutely nothing but carrying her groceries up the stairs. But an 80-year-old woman is very grateful for that, because she didn't have to do it. And because that three dollars meant so much to him, he was convinced that this woman was loaded. She probably had a zillion stashed away in a mattress in there. So he and his friend knocked on the door one day. She opened it. She knew the kid. He pushed her in and busted her skull. I mean with absolutely no feelings.

Now this little old lady's son shows up while we're there. She's in the morgue body bag laying in the hallway. I never saw the cops do this before, but they said, "We'd like to show you the body. Is this your mother?" They

•

opened up the bag right there, and he saw his mother in the uncleaned state, because they clean them up first down in the morgue. I felt really terrible for the guy.

He was an attorney, lived in Jersey, big house, big money. He had been begging his mother, "Mom, I have enough room, I can build you your own separate little apartment, attached to the house, where you can live. I'll put a kitchen and a bathroom in there, and you can come in and see your grandchildren."

She was an independent woman. She said, "No, I want to stay here."

He couldn't get her out, and that was sad. That was really sad. This guy was not a guy who had ignored his mother. He used to come around to visit with her. She paid a price with these heartless bastards. I mean this kid was cold, the way he told the story, "Yeah, I wanted the TV. She didn't have any money, so I took the TV."

I never found out what happened to the kid. I'll tell you the truth, I never cared. I look at it this way. After Willie walked out and asked me for a buck to get home, I don't care what the judges do. I'll do what I have to do to find out how the fire started and make an arrest. But what the rest of the world does with them after that is not my concern. I can't tell them to put this guy away.

I found that out very early on with Willie. I mean, he turned a completely livable building into a vacant shell, because that water came down the stairs, and those people moved out, and as soon as those people moved out the pipe strippers moved in, and they made it less habitable. Within three or four weeks that building was vacant. I think what Willie did was horrendous, and he only got six months' probation; after that I never paid attention to what they did.

With my own cases I go down for the grand jury or go to the trial, but when I'm finished testifying I get up and walk out. I'd never call the DA. Because if I called up and

•

checked on everyone, I would probably find that they didn't get what I thought they deserved. Then I'd wind up with an ulcer.

I always tell the guys, "We're not priests, so don't get yourself in an uproar if the judge says *dismissed*. You did what you had to do."

I read in the paper the other day that someone was convicted of killing somebody, and they got two years. I don't care who the human life is; it seems to be worth a little bit more than two years. Most time you do the minimum unless you're a real scuzz bucket. They keep saying that the jails are overcrowded and that sort of stuff. But if you pull certain crimes out of New York City, for instance, if you go out to Mudpuddle, Utah, burn a house, and kill somebody, you're going to do some pretty stiff time, 10, 12, 15 years. They don't worry about the jails being overcrowded. They just ship their asses to prison and that's it, good-bye.

Here in New York City it's turnstile justice. They're in and out, in and out. the DAs say, "Oh, it's the caseload. If we have to go to trial with every case we need more judges." Well, if we do, let's hire some more judges and get more DAs and build more courthouses.

I'd like to see them break Manhattan up into Manhattan north and Manhattan south with two separate prosecutors because it's still the number-one place. The number of arrests that come through the Manhattan courts is outrageous.

As for the DAs, some are very good, others are just there learning. One day they're an ADA, the next day they're a defense attorney. I have somewhat of a moral position: If you're a criminal I'm out to get you. I don't think I could turn around next week and be a private investigator trying to prove that you didn't do the crime. I just wonder how some of them can be a prosecutor one minute and then hang their shingle out and defend somebody who they know, coming in with a record a yard long,

•

is guilty. I know the Constitution says we all have the right to a trial and are presumed innocent, but I have a case here. It's one of the juveniles, and this kid doesn't stand a chance.

We're starting to get smart in this juvenile outfit. We now run mom, dad, aunts, uncles, everybody for a criminal record. We want to find out what this kid is living with.

Mama's record comes back. She's been locked up all over the place for prostitution, for drug use, for possession of stolen goods. These records are all computer printout sheets, and they fold. Mama's was about three sheets.

Papa's came back. We laid it out in the hallway, it was 23-feet long. There was his date of birth, let's say April 1 of 1950, and April 2 of 1966 when he turned 16, bam, there was his first criminal justice record. I'm sure that he took hits when he was 15, 14, 13, 8 and 9, but it never shows up on the criminal record. But his record ran 23-feet long on computer paper. I think he did 3 years at the longest stretch.

We have another one where a kid died in a fire, and they said that the kid started the fire. Our guys found out that this kid was living in an environment of crack heads. His mother was a crack head, and she's living with three brothers who are crack heads and dealing crack. We think that maybe it was the crack that started the fire and not this kid. Of course he's dead, but even mama's pinning it on him. She was stoned. She was asleep all right, but she said, "Oh, it must have been him." Well, we run everybody. The oldest brother comes out. Let's say he is 30 now, so his record was 18 feet. He took a hit every month from the time he was 16. Every single month he took a different collar for burglary, possession of burglars' tools, possession of stolen goods. Those were the crimes. He was a small time, lightweight burglar. But he didn't do anything except "time served." He did 10 days in Riker's. He did 30 days in Riker's. But he didn't do any hard time.

•

Criminal justice here sucks when you see stuff like that. The way the records are set up, it's the date of the arrest, the charge, and then the disposition, and you keep looking at this thing: "30 days, time served," "Case dismissed," "Case dismissed." I don't know what it is. I don't have any answers other than to say you need more judges.

There is one case involving a DA that springs to mind: Jimmy McSwigin, and my partner, Neil Lagatta, and I were eating in a restaurant down on Second Avenue and Eighty-fifth Street in Manhattan. Good neighborhood. We're paying the bill. We're standing at the end of the service bar giving the money to this waitress, who we were always kidding around with, and we hear *boom!* I thought it was an M-80; they were popular at the time. I look around the restaurant, and there's this guy with smoke coming around him. He's in a cashmere overcoat, three-piece suit, nice tie, manicured fingernails. I mean this guy was really well dressed. Neil says to me, "That was a gunshot."

I said, "Are you sure?"

"Yeah."

I see the manager picking up the phone, dialing 911.

"Neil, I bet you this guy is a cop. I bet you he's in his cups, and he shot his gun off."

We leave the restaurant thinking we're going to help this guy get in a cab and get the hell out of here. I say, "Hey pal."

He turns around. He has a gun in his right hand. He sticks it in our faces and he says, "Get away from me or I'll kill you."

I'll tell you, that gun looked like a cannon. I said, "Okay pal, go ahead."

This guy turns around and starts walking up Second Avenue.

Neil ducked in the doorway. I go out into the street and get behind a car. I figure, hey, this guy aimed this gun at me, so all bets are off, cop or no cop. I take out my gun and yell, "Police, stop!"

•

He turns right around. He's now outside of Jackson Hole, the hamburger joint on the corner. He shoots at me. *Boom! Boom!*

I shot. I hit him in the ankle. I'll tell you I was lucky with that shot, because as far as I was concerned the car I was behind wasn't big enough to protect me. All of a sudden it became small. I'm trying to crawl down to the curb.

That gun going off in the street has a distinctive sound. Up at the range we always fire with the ear protectors on, so you don't hear the ping. But you fire on Second Avenue with the sound resonating off the buildings, and you have no ear protection, it's a very, very unusual sound. You don't foget it.

Neil is still in the doorway, and he fires. *Boom! Boom! Boom!* I hear his three shots go off.

The guy turns the corner, and he starts running up Eighty-fifth Street towards Third Avenue with a bullet in his ankle. I thought I had hit him, because I saw his leg jerk when I fired the gun. But anyway he runs up the avenue, and I'm chasing him. It was like out of those old cowboy movies; he's running and he's firing over his shoulder. *Boom! Boom!*

I want to shoot this guy but coming down the street is this little old man with a cane, and he's got this Charlie Chaplin waddle. He's looking around hearing these shots.

The well-dressed guy empties his gun at us. Then I hear him going *click, click, click* on the empty cylinders. Your mind is racing through all of this. We run up, grab the guy, and throw him up against a car. He threw the gun away as soon as we got close to him. But we recovered it. We cuff him and pull out a six-inch hunting knife he had in his coat pocket.

After we got the guy cuffed, a sector car pulls into the block; and here's the way cops think. The cop says, "Who's the marshals?"

"Right here."

•

"Call your union representative." That's the first thing he said. He didn't say to give the Miranda warning or ask if he was searched for another gun. He says get the UFA rep down here.

Then another guy comes down the street in a T-shirt and slippers and a four-inch Smith in his right hand and a cop shield in his left. He's asking, "You the good guys?"

I said, "Yeah."

He said, "I'm sitting up at the frigging window watching this whole thing take place. I was ready to shoot somebody, but I couldn't tell who was who."

I'm saying to myself, you can get shot anywhere in this city; anybody could be shooting at you. Then I think, I hope they find all the bullets, because I shot twice, my partner shot three times, and he shot four times. That's nine rounds. I'm just praying to God that there wasn't some little old lady sitting up at the window that they're going to find, two weeks later, dead in her apartment with a stray round. They never found all the rounds. God knows where they went, but thank God, nobody else was hurt.

I'll tell you what I did. I shot the guy in the ankle, and I fatally shot a '67 Chevy. I put one right through the frigging door of the '67 Chevy he was standing by.

We lock this guy up and we run him: This was the fourth gun collar in New York City since the time he was 19 years old. He didn't do any time. "Case dismissed," "On probation," all this other stuff. His gun was stolen from Florida. He was a merchant seaman, good-paying job. He wasn't robbing the joint. He was drunk. He was in his cups, and the gun went off. He liked guns. He probably had his hand on it, rubbing it, or whatever, getting his cookies off, when it went off.

This guy is a head case. Jimmy McSwigin had to guard him, for 10 hours, down on Bellevue Hospital prison ward. Jimmy still razzes me about that. He says, "goddamn it,

•

A third-alarm fire caused by children playing with matches. Miraculously, no one was killed. (R. Athanas)

Two children rescued from the third-alarm fire. (R. Athanas)

Fire marshal James McSwigin (left) and supervising fire marshal Arthur Massett determining the point of origin of a juvenile-set fire that took the lives of three children. (FDNY)

Fire marshals William Manahan and Angelo Pisani developing leads in an investigation. It can take weeks of this work to break a case. (FDNY)

A fire marshal frisking a suspect—a dangerous point in any investigation. (H. Eiser)

Clues from these bedsprings, on which a woman's body was found, helped fire marshals William Manahan and Donald Green convict an arsonist. (FDNY)

Fire marshal William Manahan bringing in a defendant in a hotel fire. (FDNY)

An arson fire in a commercial building. Such fires often threaten the entire
block. (FDNY)

A fire marshal gathering information about an incident at the scene and time of an arson fire, before memories and cooperation fade away.
(R. Athanas)

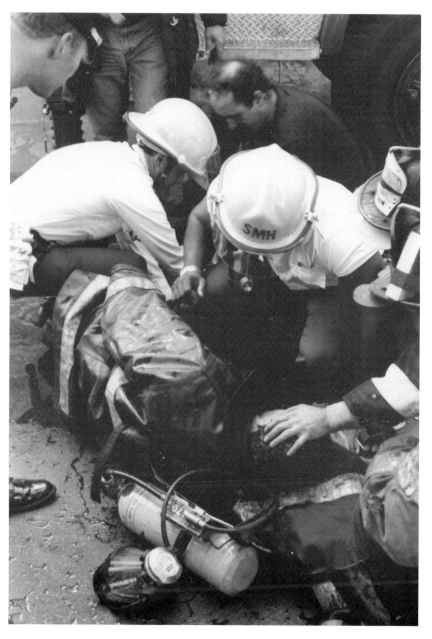

This fireman broke his leg at an arson fire in which gasoline was literally pouring out of a burning building. (R. Athanas)

A gasoline can retrieved from a suspicious fire. (R. Athanas)

A toddler being rescued from a fire. Most fire fighters will tell you that nothing is more satisfying than saving a child's life—or more terrible than losing one. (FDNY)

Fire marshal Jack Carney searching for the cause of a fatal fire. The char patterns, or "alligatoring," on the beams reveal clues about the intensity and duration of the fire. (FDNY)

Fire marshals Jack Carney and Bill Mulhall reenact a night shootout in Brooklyn. The perpetrator, who had fired into a crowd and wounded one person, was preparing to shoot again. He ended up paralyzed from the waist down after being shot by officers. (FDNY)

Fire marshals John Knox (right) and Bill Maxwell arresting a suspect in an arson-homicide. (FDNY)

The top-floor rear: A favorite location for arson-for-profit fires. (H. Eiser)

Deputy chief fire marshal Tom Clarke training with a shotgun, the assault weapon carried when heavy fire power has to be used. (P. Micheels)

Tarps cover the victims of the Happy Land fire. (S. Spak)

The largest case of mass murder in New York City brings Chief Fire Marshal John Regan (right), Tom Mulvey (center), and Regan's soon-to-be successor John Stickevers to the scene. (R. P. Cook, Bronx Press Review)

An arson fire in the Grace United Methodist Church. To the local junkies who set the blaze, believing that the church had informed on them to the police, no place was sacred. (H. Eiser)

you got a medal out of this, and I had to stand watch on this guy."

The next day I had to go down to criminal court in Manhattan and this woman ADA—first thing she does is advise me of my rights. I said, "Time out, what's going on?"

She said, "That's standard procedure."

I pick up the phone and call the chief fire marshal.

He says, "Yeah, that's standard procedure. When a police officer shoots someone he has to waive his rights and give his testimony as to what happened."

I waived my rights, and she gives me the Miranda. I sign the waiver. Now she talks to me about what happened.

We had a witness—talk about a stand-up guy. A civilian is standing across the street, he hears the gunshot, and he hears someone yelling *police*, he sees me turn the corner, and his immediate reaction was to get behind a car. "But," he said, "I stood in the doorway and watched, because I figured that this man was going to need someone to testify on his behalf." He came right up to me and gave me his name and address. He went to the grand jury and testified like he was a video camera. He gave a vivid description of this incident. I didn't realize it, but I had said, "Police, stop!" about half a dozen times.

The seaman's defense was that he didn't know we were police officers.

After the ADA gets this whole story and the arrest papers saying that I'm charging him with attempted murder of a police officer, that's when she tells me, "Well, I don't really think his intent was to murder you; his intent was just to get away from you." So she knocked it down to some sort of escape. I think it was escape one and attempted assault and all this other crap. He got seven years out of it. I mean he got a pretty good sentence, but they sort of let you know where you stand in the scheme of

•

things when they knock it down from attempted murder. I swear to God I thought he was trying to murder me. He shot four times at us; to me, that is attempted murder. But she said no, his intent was just to get away from you. If he happened to kill you in the offing, so be it.

I was just amazed. First of all, at that point, I was sort of numb, because this happened at about 11:00 at night and we were up all night with the reports and then went right into court the next day, but the first thing you get when you walk in the door is an official at the prosecutor's office giving you a Miranda warning. You're saying why the hell are they coming after me. Then she's trying to explain to me that she doesn't think that this guy's intent was to murder me. I said, "This is ridiculous." I did what I had to do, and I can't do any more than arrest the guy. That's it. Once he's arrested I'm finished. She chose to go the route that she did, and I just walked away from it. I shook my head. I didn't understand it, but I said, "Hey listen, that's the way the cookie crumbles." I think I saw him a few more times. He made bail after he got out of the hospital.

I'll tell you, the ADA was feisty. When he was arrested, because he went to the hospital, all his possessions were confiscated. He's a marine engineer, and apparently the way they get paid is they get a voucher at the end of their trip, and they go into the Seamen's Hall and turn it in for cash. He had a whole bunch of these in his wallet that he hadn't cashed in. He was owed a lot of money. She took these, and she took his credit card and his seaman's card. Her argument before the court was this guy has got international travel privileges, he can just get on a boat and be gone and never come back again. I don't want the guy leaving town. So she took all his means of getting away and held onto them, and that brought me into court more times than the actual shooting charge, because his attorney was always going back into court trying to get her to release the pay vouchers, because the guy needed money to pay

•

his rent, and of course to pay his attorney. Every time he needed money, he would go to his attorney. They would make a court date to show up in court and ask for more money. She'd ease up and give him another pay voucher, but he never left town. He did go to jail.

He was a big Irish guy who lived on Ninety-first Street and Madison Avenue. I think he was a graduate of Cardinal Hayes, which is a prestigious Catholic high school. If you looked at him you'd swear that guy's a detective, dressed to the nines. Who would have figured it.

I was convinced that I was going to help this guy out. Instead he's trying to shoot me. Strange thing. That one gunfight was enough for me. I don't want to have to see that again. I mean for a month or so after every time I heard a car backfire it made me jump.

There are different feelings when you lock somebody up. With Willie I felt very, very good. I was really glad that I was taking this mutt, who inconvenienced all these hard-working people, and locking his ass up.

That kid with the pregnant girlfriend, I really felt bad that he had to go. I mean there are some guys that are just dirty nasty criminals all right; there are other people, who got caught up in whatever their personal situation was and saw this fire as a way out, that I felt bad for.

I locked up a guy living in a West Side hotel who made a bunch of fires in there. He was making fires to get back at people who had busted his chops. He would make them against their doors. I didn't feel bad at all locking him up.

A few years ago they improved the system for arresting someone. Now when you lock a guy up, you take him to a precinct. There are police department forms that you fill out: The on-line booking sheets that you fill out to describe the individual's pedigree and then the crime that you're charging him with, the details of what went on. Then you take him to central booking, where you turn him over to the police who are in charge of jailing him in there. They log the prisoner in, and he's fingerprinted. You go upstairs

•

and tell your story to the assistant district attorney. He translates it into the legalese and types up an affidavit. You then go down to a sergeant in the PD and swear to the truth in that affidavit. Then you leave. Your prisoner is in jail, the affidavit sworn and signed by you is there with a representative of the police department, and when it's time for that prisoner to be arraigned before the judge the police take the guy out from the pen, and the judge has your sworn affidavit that this is what the guy did.

In the old days before they had prearraignment, you did all of that yourself. You put the guy in the cell, and you filled out all the paperwork, and then you sat and waited for him to be arraigned, and then you went and stood before the judge, and the judge reads your affidavit, the bailiff swears you to the truth of that affidavit, and then they make their decisions: whether or not they're going to plea bargain, cut him loose, hold him for a trial or whatever. But you had to physically be there. Nowadays you don't have to for felonies. And arson is a felony. For misdemeanors you still have to stand arraignment, but we don't really get involved in that, because all of the arsons are felons. So it cuts down our in-court time a lot.

The next step in the proces is when you get called down to testify before the grand jury. It's nerve-racking the first time you go, because you walk in and 26 people are sitting there, and you're sworn in before them, and the ADA is in there, and he leads you through a description of the crime. Then you leave the room, and these 26 people vote on whether or not the individual should be indicted for a felony. Once they hand out an indictment he goes to trial, if it ever gets that far, and then you get called in for the trial—the Perry Mason thing where you sit in court and you testify before a judge and get cross-examined and all of that stuff. But I'll tell you, in 11 years I have done marshaling I think I've been to trial 3 times. Most of the stuff is handled in plea-bargaining.

At this point in my career I have joined the juvenile unit.

•

It's interesting, and it's very, very good. We're finding many of these kids are absolutely living in hell. Mama is a crack head, Daddy ain't around, the kids are left alone, they don't have food, they don't have clothes, and they make fires to get out of it. I think they're brutalized, not always physically, but I think they're brutalized mentally. Take an eight-year-old kid in an apartment that has no electricity and no heat in the middle of winter, and that whole apartment is pitch black, and you leave that kid in there all night long while you're out on the street trying to get some dope. I think that kid's been brutalized.

We had one case of an eight-year-old girl. She's in the apartment, and her mother is out. By her mother's own admission, she just came back from shoplifting to get food, because she'd used all her welfare money to buy crack. They had a dog in the apartment that the mother calls "bitch," that is only controllable by her. Her little girl is terrorized by the dog which the mother keeps locked up in the bathroom. The girl cannot go to the bathroom, because this dog is in there. The mother rented the girl's room out to hookers for 30 minutes at a time. The mother is all upset one afternoon because she can't find her earring. This kid lit a candle, when mama was out shoplifting, and climbed underneath the bed to look for the earring. The candle set the bottom of the mattress on fire, and the rest is history. As bad a mother as she was, the kid was there trying to please her, trying to find the earring for this piece of garbage.

We called SSC, Special Services for Children. They're the child protective services in New York. They took that kid away and wouldn't give her back to the mother.

We get a kid in the first grade, seven years old. He set fire to books, in his desk, in his classroom. We talk to the kid, and we make a referral to the community mental health agency. Two days later I get a telephone call from the therapist, and this kid is giving himself and five or six others up as being victims of sex abuse. The local pillar of

•

the community had molested him and the other boys. We turned it over to the police sex crimes unit. They locked the guy up. It turned out he had been previously arrested in Queens for child pornography.

We get kids who set "the cry for help" fires whose mothers are prostitutes, the fathers are either drug addicts or drug dealers, or both, or the parents are alcoholics. These kids we turn over to mental health. You talk to them, and they're very, very smart kids. But, they're just trying to get out of a situation that they can't deal with, and they use fire to call attention to themselves. I don't know if they do it consciously, because I don't think a seven-year-old kid could be that devious to say, "Well, I'll set this fire so someone will come along and see what the problem is and take care of me." I think they're just burning this place up to get rid of it.

We had a fire set last week where the kid made a fire in his closet. Just to give you some background, the mother is an alcoholic. The father, most of the time is not around, but when he does show up he's usually in his cups and gets into a fight with the mother and busts up all the furniture in the house.

We asked the kid for three wishes. He didn't have any wishes. Here's a seven-year-old kid who has just given up. His outlook on life is why wish for anything, I can't get anything anyway.

I go home to suburbanville, 60 miles away, and I hear my kids complaining because they have to unload the dishwasher, and I say, "Come on guys, you want to see what life is really about?"

Jack Bowens and I walk up to an apartment, and there was no lock on the door. We pushed the door, and it opened on its hinges. We walk in. The mother looks like she was high. There was an unidentified male sitting in the kitchen draped over the table. The bedroom, well what acted as the bedroom, living room, den, conservatory and

•

library, had a big bed in the middle of it. The sheets on the bed were actually charcoal gray from dirt. They had a fire in the apartment. The back bedroom windows are the plastic put up by fire salvage. The walls are ripped down, the ceilings are ripped down. No attempt was made to clean the place up; there was just fire debris in this room. The door on this room was burnt off. The bathroom consisted of a sink hanging off the wall and a couple of five-gallon pails of water. Where the toilet should have been was the old lead band and the hole for the pipe going through the floor. If they had to use the toilet they used this hole in the floor and then poured water down. The SSC worker pulled the kids out of school and would not let them back into that apartment.

In one of our early fires in the juvenile unit an infant died. I'm going to show you the irony of this thing. There were three surviving kids. One was a four year old who had AIDS-Related Complex (ARC). The kid probably has full-blown AIDS by now, because this was a year ago. Mom is a junkie.

Someone said the reason they took these kids right away is that she had a prior fatal fire. So I go through the reports and it turns out that I worked on that prior fire seven years ago.

She didn't actually have the fire in her apartment. But two lesbians who lived in the apartment underneath hers had gotten into a fight, a lovers' dispute. One set fire to the bedroom. This was at 1:00 in the morning—this fire killed two of the kids upstairs who were left alone by this mother. The kids were two and five years old. At 1:00 in the morning, their mother was out doing junk.

The lesbian was locked up by the PD. China was her first name. I'll always remember that. As soon as I saw the report, and I spotted the name China I said, "Oh shit, I worked on that fire with my brother-in-law" (who was a marshal then).

•

Now she has a whole new family, and she loses another kid in a fire. The four year old with ARC was the fire starter. He was playing with matches.

We also had a kid who was trying to burn his school. He burned a satanic star in the middle of the schoolyard. Then he trailed the gasoline in this serpentine course throughout the schoolyard. You could see where the tar had melted, right up to the door of the school. It turned out he was school phobic. He was terrorized at the thought of having to go to a new school. He was living in a house where the mother was a manic-depressive, who refused medication. The father cooperated initially. The doctors told him there was something wrong with his kid. The kid was diagnosed as a pyro. But then the father said, "This is too much bother. There's nothing wrong with my kids. I'll handle it myself." He was a hard-working, middle class guy.

But many of the kids you see are living in shit. It's an interesting thing, when you go into the apartment, and there's no toys. There will be three or four kids there, but you won't find a toy truck or anything, just the TV, that's it. It's sad when you look at them. You wonder how they survive as long as they do. Also you can understand why they turn out like they do as adults, because they had shit when they were kids, and when they get old enough to get what they want, they take what they want and that's it.

Maybe we can do something by getting them into some sort of therapy. I don't know if it will be helpful to all of them, but I'm sure we're going to do something positive for some of them. Now that's what makes this thing kind of interesting; it's a step in the right direction. Being the typical fire marshal is punitive, because you're coming up to cuff somebody and take away his freedom. Here you're freeing somebody up from being stuck in the cesspool that they're in, so they can see the light of day and get out of it. In that respect it's a breath of fresh air.

•

Because to tell you the truth, I was getting bored with being a marshal; going out, seeing this burned-out building, let's go lock this guy up. Maybe that's something that's part and parcel of me; I get bored easy. But this is good. This is like mom's apple pie. You can't be against this.

But the marshals are affected by it, especially the family men. There for the grace of God go you and I. They feel for these people. Patty Farrell and Jack Bowens have been around here as long as I have—11 years. They've seen all of the garbage, and I thought they would have been hardened off by it, but they're not.

We'll sit here and discuss cases and B.S. amongst ourselves; and the common phrase that comes out is, "Boy, these kids really got it tough." We make light of some of them. Like I was in an apartment last week and this kid left footprints everywhere including the ceiling. This was a hyperactive kid. He was like Gerald McBoingboing. He was off the wall, and we laugh about it. If we don't laugh about it, if we keep everything in, I think we would probably go off like a Roman candle. We have to talk about it, and we have to make light of some of this stuff, like the dog named "bitch" and that sort of thing. We realize the seriousness of it. But it's gallows humor.

This program is the best thing to come down the pike in this office in a long time. It's a new way of dealing with a problem that we've been walking away from for years and years. When we were doing those early statistics we'd come up with 3 kids that made 38 fires. I would have liked to have been able to sit down and go over the detail of fires with these kids and then run them through the property-insurance loss register and see how many dollars these kids were burning up. They're making some big fires.

There was a second alarm the other day right up the block on the Grand Concourse that took off the top floor. A five-year-old kid took matches into the closet, that's all.

•

He just wanted to see what it was like inside the closet. The match burned his finger. He threw it on the floor. It caught on some clothing. The rest is history. He wasn't a bad kid, but he caused a lot of damage there. We did a little educational intervention with him, showed him our pictures. "Fire is a tool" is the analogy that we use. Hopefully he won't do it again. From the stats we've been pretty successful with it.

We have had some recidivist behavior, because there's been a breakdown in the system. Either mama is not taking the kids to the doctor, or daddy has shown up and said, "This is bullshit. You don't have to take the kid to the doctor." Whatever, the kid is not getting the help that he should be getting, and that's what happened with the three recidivist cases that we had. And the way we found out about one of them is that mama calls us and says, "Hey listen, I think you should come over here and talk to my kid again. I caught him playing with matches."

I said, "Well, have you been taking him to the doctor?"

"Oh, no, we haven't got time."

We go over there, and we talk to the kid again, and we tell the woman, "Get the kid back into treatment."

It's not the kid's fault. A seven-year-old kid can't make decisions for himself. I keep saying 7—there are 10 year olds that can't make decisions. So it's mama or papa that's at fault.

I think, without getting too burned out, I could do another 10 years here. And that would give me 30 years in the department. As long as there's some positive results—and we're seeing positive results with the juvenile unit, based on the feedback that I'm getting from the therapist.

My father was chief of the fire department in 1963. He was there in "the days of iron men and wooden ladders." He would always say, "Why don't you get an honest job and get back in the firehouse?" But when I told him about

•

this juvenile unit he said, "That's great. That's the kind of stuff that should have been done years ago." And this was from a man who is 88 years old and rather set in his ways and ideas about the fire department. So, if I have his approval on it then I know I am doing the right thing. And I'll do it even though Jim McSwigin is the director.

•

•

# 5

# FIRE MARSHAL

## JOHN KNOX

\ came into the fire department in February 1960, after spending five years in the Marine Corps. In the fire department I met the same type of guys that I knew in the Corps. They were men that you could really rely on. You may have differed with them intellectually, politically, or ethnically, but when a job had to be done they always got it done. I enjoy working in a team. I found my niche.

The fire marshals had a mystique to me. I became a marshal in 1964. It was the beginning of the fire storm that raged across the city into the mid-1970s. Brownsville was starting to burn, the Bronx was heating up, and we only had 62 fire marshals.

One of the earliest investigations I was involved in was used as background for the movie *Cruising*. Seven or eight murders of homosexual men were committed over a relatively short period of time. They were muti-    lated and killed, and then their private parts were set on fire. In most instances the fire extended to the room they were in. The fire was probably

•

used to obliterate the evidence of the crime. The murders all took place in the Greenwich Village neighborhood of Manhattan.

At one point, a body was found floating in the Hudson River off Greenwich Village, and there was an outrage from the gay community.

The gay community's spokesman had a meeting with the police and the fire department. The gays felt that we were not doing enough to find the murderer.

This was the first time that I met our former mayor, Ed Koch. Back then he was the attorney for an organization called the Matachine Society, which was a group of homosexual professional men. At that time, Koch was a member of the Village Independent Democrats, in Greenwich Village, that was trying to throw out Carmine De Sapio.

A task force was formed between the Bureau of Fire Investigation and the NYPD. I was one of the investigators on the task force, because I had covered some of the original DOAs.

I had worked on one that at first we didn't think there was anything suspicious about. We thought a guy fell asleep while he was smoking. That was over on Horatio Street.

We found two other bodies in a building that was right across from the old firehouse on Varick Street that was used in the film *Ghostbusters*. It appeared that these two men had been strangled, but one of them was also stabbed. Their testicles and penises were burned. They were around 28 or 29 years old.

We started comparing notes, and we found out there were several other people similarly caught in a fire.

We worked on this case for about two years. We had four teams of fire marshals going around to these gay bars showing pictures of people that were given to us, because they weren't seen anymore on the gay scene. We didn't know how many people disappeared. At that time, the

•

homosexual places were more clandestine than they are now.

We came up with many peripheral people who were involved with these homosexuals, but they were living supposedly straight lives. Some of them were married, had children, and were very successful.

One of the guys that we were involved with was a bouncer for some of these homosexual places. His name was Skull Murphy. He was also a professional wrestler. Skull was working for Matty "the Horse" Ianiello, who is now in prison. Matty was a member of either the Luchese or the Columbo organized-crime family and was in control of these gay bars.

Skull Murphy knew all these homosexuals—he was a homosexual himself. Skull had worked in hotel security, but he was involved in shaking down people. A guy from out of town staying at the hotel would want a boy. Skull would supply the boy, but because he was security he'd later bust in and take pictures. Then he would extort money from the guy.

Skull went to jail for that in 1964. He was out in 1967; that's when this case started. Skull died last year.

We were put onto a guy who was a chicken hawk—a guy who recruits young boys for homosexuals. He was involved in something in Jersey, and he said that he knew where five or six homosexuals were buried out there. But he turned out to be a fart in the wind. We got the Jersey City cops to dig up this dump under the Pulaski Skyway, but they didn't come up with anything.

However, we did get the Feds to hook up with this guy, and they were able to break up this ring that was transporting the boys from state to state.

The murder case was going nowhere. Then one evening around midnight, a cop was walking a beat on Fourteenth Street, when a guy comes running out of a building and bumps into him. The cop grabs this guy who smells like smoke. Then the cop looks up and sees smoke coming out

•

of the building. He handcuffs the guy and brings him back into the building, where the cop finds another man tied up and burning.

The guy who was caught running out of the building never admitted to anything else but that fire and the assault on that man, but after he was locked up all the killing stopped.


I was working with a guy named Bill Gaynor, who was a civilian fire marshal. We had a fire in a bar called the Crazy Baby. It was on Amsterdam Avenue in the seventies, or the eighties, near the Museum of Natural History.

Above the bar were five stories of people living there, Puerto Rican people—an occupied building. This torch was going to set fire to the whole fucking building, but a cop was going past the place, at 4:00 in the morning, and he saw this fire starting and called the fire department.

The fire department was only a few blocks away. They knocked the fire down really quick, and there was no damage. They found this incendiary device almost intact.

When you rang in on the telephone this device was activated, producing a spark, and igniting a container of naphtha. Somebody told them that if the fire started there wouldn't be any residue, because the naphtha would explode.

Now the guy calls up again at 8:00 on Sunday morning. I picked up the phone, and he hung up. He's calling the bar, because he knows if the phone rings the place is still there.

Eventually we call up the owner. He comes down, and he starts his shit. He is a very prominent lawyer, and he was appalled that anybody would want to set his bar on fire. He goes on to tell us that he has the only key to the place. He also did something that he'd never done before, that is, he locked the place up with this other guy, his

•

bartender. We asked him who the bartender was that was with him.

We went to the place where this bartender lived and snapped him before he could talk to anybody. He thought that he was going to come in on Monday and find the place burned. We told him what happened and that the lawyer ratted him out. He then ratted the lawyer out. It was very easy.

The lawyer went to jail for five years, but he was out in a year and a half. He was also disbarred.


My longest-tenured partners have been black. I worked with Aubrey Nelson for 10 years and with Jimmy Callender for 3.

Aubrey lived in Harlem all his life. He had two years of college. He was an athlete at De Witt Clinton High School in the Bronx, and he was a paddleball champion.

When I was working with him, I wore double-breasted jackets, and he wore a dashiki with sandals, his hair braided, and a pocketbook. The only thing he didn't have was a bone in his nose.

So, I got into the firehouses, and he would be outside fighting with the firemen, because they wouldn't let him in. When we walked into a precinct the cops would tell me, "Your prisoner has to be rear cuffed."

"That's not my prisoner, that's my partner."

One night we had to go to Harlem to see this Reverend Dempsey. They used to call him the pistol-packing reverend. He had the upper Park Avenue Baptist church, on Park Avenue and 125th Street.

Well, before we got up there Aubrey said to me, "I don't want you fucking with these niggers tonight. I got a paddleball championship to go to tomorrow, and I don't even want you looking at them."

At that time, there was a group of blacks up on 126th Street, between Park and Madison avenues. They were

•

called the Five Percenters. They used to wear bones in their ears and in their nose.

Well, we get to where we are going, and Aubrey sees these guys with the bones in the nose, and he says, "John, remember what I told you, don't be fucking with these niggers. We're going to go talk to the reverend."

I get out of the car, and I meet three of these guys with the bones in their noses, and they start lipping on me, so, I get on them. The next thing I know we're rolling in the street with all of them. The streets are filthy.

This is when they had the garbage strike in 1969. It was 105 degrees. It had been over 100 for 3 days in a row, and the city was burning.

I got this guy, and I'm trying to pull this bone out of his nose, because I told him I was going to pull it out.

The cops are pissed at us because they had to come out, because there is a riot on the street.

I did get the bone out of his nose, and I put a couple of lumps on his head. He got locked up for assaulting me.

We had the Harlem riots when Martin Luther King was killed. They were looting the stores on 125th Street. So, Aubrey and I are standing on Fifth Avenue and 124th Street when a guy goes by with a shopping bag. Aubrey says, "Hey, this guy's got four Molotov cocktails in his bag." So, he goes around the corner after the guy.

The guy says, "Brother, look out," and he throws this Molotov cocktail through the window of this supermarket. Aubrey says to him, "Come on, this way." So when the guy comes back around the corner we grab him. We didn't have our car with us, because we were told to leave it at Mt. Morris Park. So, we had to take the Fifth Avenue bus, with this guy in handcuffs, to the 26th precinct.

I'm working with Wesley Powell, who is a black guy, and Enrique Henry Estell. Henry has a heavy Spanish accent. He was captured by the Chinese during the first

•

part of the Korean War, and was a prisoner for two years in Manchuria. He speaks fluent Chinese, but you never knew whether he is speaking Chinese or Spanish.

We get called to Capox Street up in Riverdale in the, Bronx. The fire department responded to four fires in this building, but the super tells me there were about eight fires over the past week. He tells me he knows who made them. He said, "It's this woman. I caught her one time when she set a fire in the incinerator room."

I asked, "Why is she setting fires?"

"I don't know, I think she's drunk all the time."

So, we knock on her apartment door, and this well-dressed young woman, about 28 years old, opens the door. She has this little boy baby in her hand, and she is starting to cry.

I said, "Let's talk." We all walk into the apartment. I ask her, "Why did you set those fires?"

"I don't know. I'm always alone, and my husband is always working."

"What does your husband do?"

"He works for the CIA." I figure this is part of her aberration, but she shows me something that says he works for the Central Intelligence Agency. "He works on Broadway. They have a building over there where the old Paramount Theater used to be. He's a research specialist."

Then she admits to all the fires, and she gives me a signed confession. She calls her lawyer, but she can't get help, because it is 10:00 at night, on the 17th or 18th of December.

I'm looking at this little baby, which is about a year old, and I am thinking if I lock this woman up I have to take this kid to the foundling hospital, but first I have to take him to get examined. So, I said, "Where is your husband?"

"He's at this number."

I call him up and I say, "Hello, I'm arresting your wife." He said, "What?"

•

I said, "Yeah, your wife. You had better come home, and I'll explain it to you when you get here."

It took him two hours to get home. When he walks in the door he sees a guy who looks Italian, a black guy, and a Puerto Rican sitting on his couch, and he says, "What's going on here?"

"Your wife is under arrest. I called you so that you can watch the kid."

He says, "Look, I'm an Irish-American, and my family has contributed more to the growth of the United States," blah, blah. Then he starts to get nasty with me.

I said, "I don't know what you're about, but your wife's under arrest. I got room in the car for you too."

He starts telling me that we aren't taking any one of them out of there. I said, "Your wife told me you worked for the CIA. I don't know what that means, but if you have a gun on you. . . ."

"I have no weapon on me."

I said, "I'm a little annoyed now. I'm trying to do you a service. I don't want to have your kid in a foundling hospital. Your wife's a drunk, and you seem to be another type of scumbag."

Now he wants to talk to the supervisor. I said, "You don't want to talk to him, he's a bigger prick than I am, and he's Jewish." I didn't tell him I was Jewish.

He said, "Let me speak to him." So, I get the supervisor on the phone. I give him the phone, and he goes through this whole thing about him being Irish-American, and he's entitled to a certain amount of courtesy, and all that. Then he hangs up and he says, "You're right, he's a bigger prick than you."

Well, we arrested her, but they gave her a desk appearance summons, which means that she had to appear in court at a later date. Then we took her down to Columbia Presbyterian Hospital, and they put her in the psychiatric unit.

•

I never heard of the case again. I just felt sorry for the woman. When I met him I could see why she was that way. She was going through all kinds of things, and the fire setting was an attention-getting mechanism. The firemen would come and talk to her.


I have locked up over 300 people. I don't remember all of them, but they all remember me.

I was walking with my daughter and my ex-wife on Broadway when a black guy comes up to me, and says, "Remember me?"

At first, I didn't, after he spoke a little more I remembered him. He was a pimp that beat the shit out of his girlfriend, tied her up, locked her in a closet and set her apartment on fire.

She was in critical condition in the hospital when we came to the fire. We didn't know that she had been removed from the apartment. This was in midtown Manhattan, over on Forty-eighth Street, between Ninth and Tenth avenues. The cop on the scene told me this woman, she was dumped in a closet. I said, "Is she dead?"

"No, I think she was alive."

So, we go over to Roosevelt Hospital. She tells us the story about this guy. She is really messed up. I didn't think she was going to make it. She was badly burned, but she was very lucid. I asked, "Where does he live?"

"He's in the Aberdeen Hotel on Twenty-ninth Street."

While we're there, he calls up to find out about her condition. So I told the nurse: keep him on the phone, we're going to go over there.

Aubrey and I go up to his room, and we tell him to open the door. When he opens the door he is still on the phone talking to her at that hospital, and he's in his underwear. It's cold. Now he was a tall, skinny guy about 6 feet tall, and weighing about 130 pounds, and he's in his underwear.

•

I said, "Okay, put that phone down." We didn't have our guns out. He makes a dive for the bed. I grabbed his nuts. Bing-bang, I pull him down. He's fighting, but we got the cuffs on him. Then I said to Aubrey, "Fuck him, let's take him like he is." Usually we let them get dressed, but this guy didn't deserve a break.

In 1979 I was transferred to Brooklyn to work with Chief Tony Romero. In 1980 we had a series of fires in a building out in Rockaway that was allegedly being terrorized by a synagogue. The synagogue wanted the tenants out of the building. So, we had guys staking out the place on Empire Avenue.

They sent me out there, because this was an orthodox synagogue, and I'm Jewish. They sent Jimmy Callender there because he's black, and some of the people being terrorized were black.

The first thing I see when I get there is this guy, by the name of Kenny Aska. I knew him to be involved with the black coalition to get jobs for blacks in construction. They were shaking down construction people in the Bronx. I had locked him up for extortion and attempt at arson. I wondered, what was this guy doing out here in Queens with a synagogue? Now when he saw me he took off, but I got his number on his license.

We first did the paper chase on the apartment house. We found that the building was sold six times in less than a year. The name of the corporation that was the holding company was Golem Realty. Golem means *dummy* in Yiddish. So, we went to Golem's office in Manhattan. We found out that one of the officers of this corporation was this girl named Sylvia, but in reality she was a secretary.

The synagogue sold the building to Golem. Golem sold it to a guy named Brown. Brown sold it to this other holding company. The holding company sold it back to Golem. According to the U.S. Attorney it was a legal transaction,

•

but every time they sold the building the price went up, and every time they had fires they collected money. They had 12 fires in 1 year, but they never did any repairs on the building. The super was setting fire to the building.

Aska was the whip. He brought in one guy from the Bronx to be the super.

At the same time, this friend of Aska, one James Blackwell, was having trouble with his Hispanic girlfriend, Louise Gomez. So, we found out where she was, and Jimmy Callender and I snatched her. She blew the whistle on these guys. She even told us about a couple of murders they committed.

Louise Gomez said that James Blackwell had shown her where they had murdered this one super from Brooklyn, and that she got off on it. He told her how he and Kenny beat this guy to death because he was shaking him down. The guy supposedly wanted more money, and to become part of the partnership. We never verified this. We know the guy was missing, but he was never reported missing. His body never turned up.

We also think they locked a young Jew up in one of those tenements on Davidson Avenue, in the Bronx. It's very unlikely to see a little Hasidic Jewish kid from Borough Park up in Davidson Avenue at 4:00 in the morning in one of these vacant apartments. They locked the door from the outside with a dead lock, and he couldn't get out. He burned to death. The guy lived for about a day, but he never gave anybody up.

Aska and Blackwell were also working for a group of landlords in Brooklyn who had their buildings burned.

One guy owned 200 slum buildings in East New York and Brownsville, off St. John's, Rochester, and Rockaway avenues. These guys burned the whole thing up.

They were buying buildings for practically nothing from the banks. They burned anywhere from 500 to 1000 buildings. It's mind-boggling. This went on for 10 years.

What took them so long to track these guys was that

•

they were in four boroughs, Manhattan, the Bronx, Brooklyn, and Queens.

In all, from the time we startd with the synagogue till
we finished four years later, we identified 39 principal
perpetrators. None of them was an actual torch. However,
all of the people we were getting were lawyers. The torches
gave them up, because these men were such lowlives that
they would stiff the torches on their fee.

One of the torches, Ramon Ayala, was a guy I locked
up about three or four times in Brooklyn. He turned on
some of these people.

One guy he ratted out was Joe Bald. Joe then ratted out
everybody else in this arson-for-profit scam. Joe Bald
didn't start any fires. He paid people to set fires. Sometimes he was a middleman, and sometimes he was the
guy who owned the property. He worked directly with
Kenny Aska and James Blackwell. The torches answered
to James Blackwell.

Then the Brooklyn DA's office tried to set up my partner
and me. They figured that we were involved in the arson
group. Well, when you're working on an investigation you
don't go to the district attorney unless you have something
substantial to indict them on.

This guy Ayala calls me up and says that he wants to
talk to me. I want to arrest him, but he's already a protected
witness by the district attorney, but I don't know this. He
wants to meet me someplace out in Brooklyn. I said, "What
the hell does this guy want? To knock me off or something?" But we meet out by Jamaica Bay. He's talking to
me, and he's got a wire on. The assistant district attorney
wired him.

Ayala is trying to get me into a conversation as if I'm
involved with him. My first instinct was to beat the shit
out of him, but I figured something was wrong, so I just
left.

Later the ADA said to me, "You know, you have got a
sixth sense, but I'm going to get you. Knox, I know you're

•

involved with these fucks." Needless to say, he never got me, because I was never involved.

I survived. I don't like to blow my own whistle, but I've maintained a certain amount of integrity throughout my whole life. I don't sell anybody, and I don't sell anything. I do a job because I enjoy it.

I was surprised by the attitude of the district attorney. I was appalled, because they were more interested in what notoriety they could get, rather than the effect that these people had on the community.

Then a year later, Eugene Gold, the Brooklyn district attorney, got arrested for molesting that young girl down in Tennessee.

Most of the cases fell apart. I believe it was because people in the district attroney's office were somehow involved.

Then it went to the feds, the feds play a different kind of ball. Larry Ruggiero was the eastern district U.S. attorney, at the time, and he took the cases.

The ATF got involved. ATF made 40 cases. We worked with agents Frank DeNapoli and Chris Behan for about a year and one half.

Kenny Aska, James Blackwell, and Joe Bald were indicted for the Rockaway fires. Joe Bald was trying to escape, but we bagged him right away; that's when he started ratting everybody out.

In fact, out of all the people that were involved in this, Joe Bald, a white Jew, did more time than everybody. Altogether, he did almost nine years in federal prison, but most of his time was spent at the Manhattan Corrections Center, which is like going to a country club.

What happens when you get guys that have never been involved with law enforcement before is that they don't want to go to jail. An old Jewish guy in jail doesn't do very well. They'll sentence them to hard time in prison with people who are young and who will terrorize them. So, they start ratting each other out, and the feds start

•

cutting deals with them, because they're looking to get the bigger guy.

The defense attorneys for these people were saying that the property that was set on fire didn't mean anything because it was vacant apartments. However, there were people living in the building, and we could substantiate it by bringing them into court.

The most significant thing that made our cases was the firemen. The juries were impressed by the amount of firemen that were hurt in these fires. Over 10 years a couple of hundred firemen were injured.

One of the buildings that was constantly set on fire collapsed on four companies up in the Bronx, and put a captain and five or six firemen out of the job.

So, the jury convicted these guys.

Then the corporation council tried to use the RICO statutes against some of the defendants, but the federal judge threw it out. He said the City took too long to bring the charges. When we told him we had very few marshals when the case began, he said that the City has the responsibility of hiring more people, because they have the right to levy taxes to get more men. If they have a problem it's their duty to correct it. The City did not act responsibly to put more men on this when it was necessary.

Jimmy Callender and I worked together on this case for three years, then he was promoted to supervisor, but I continued to work on it with other guys.

In the interim we got involved in a movie project based on this case. A guy saying he was from Universal Studios came and wrote a script, but it never got made. I realized how much integrity these guys have, and I now know what a producer does, and what a director does. This guy couldn't deliver. One time we were in Sal Anthony's Restaurant, on Irving Place in Manhattan, and he was sucking up black beauties off the table. The guy was a complete degenerate.

Aska and Blackwell were fugitives for six years. I just

•

apprehended Blackwell last year. The feds called me up and said, "We've gotten a lead on Jimmy Blackwell, do you want a piece of him?" So, I got guys from my unit, the Manhattan base, and we staked out the housing project on Alexander Avenue, in the Bronx, for three days.

The feds knew they owed me one, because these guys came to turn themselves in, and they tried to cut a deal with them. I just happened to be in the federal courthouse down on Tillary Street, when I saw these two fucks. I almost went for my gun. I said, "What are these guys doing?"

"Hey John, take it easy, they have their suitcases with them. They are going to work to rat somebody out." They were both facing 12 years of state-prison time.

I said, "I've been looking for them for two years, and now they walk in here, because you guys are cutting deals that I can't cut. I don't want these guys leaving the building," but they left, and that was in 1986.

Subsequently Aska got caught doing something else, and he's doing 12 years in prison. He won't be out until after the turn of the century.

The damage that all these people did to the community was tremendous, but for me this was the most rewarding case, because I went after these people and I got them. I grew up in the city, and I know how to step on a guy's balls.

This was the culmination of everything because these two guys caused more misery than anybody I ever met in my life. They were responsible in the 1970s for the burning of a good 150 buildings in Highbridge section of the Bronx, on University Avenue, Shakespeare Avenue, Ogdin Avenue, Nelson Avenue, Jerome Avenue, Townsend Avenue, and Grant. It is very vivid in my mind, because I was working up there and I remember being called to those buildings three and four times a night.

One time they set another fire when I was in a building on Shakespeare Avenue. It was an H-shaped, six-story

•

building. We were on the fifth floor. The fire we were investigating was in a vacant apartment. We were interviewing the people who lived in the apartment next door, when we heard screaming in the building. It's 2:00 in the morning, and there are no lights in the hall. The only lights they have are in the individual apartments.

This building originally housed 36 families, but now there were only 10 families living there, because over a period of 5 or 6 months there had been 20 fires, some of them major alarms. As long as the people weren't directly in the path of the water coming down, they were still there.

Somebody had set a fire in a vacant apartment on the third floor. As we ran down we could smell the gasoline. We had to get on the Handie-talkie and call the companies back.

The first thing I felt was anger. I said, "Son of a bitch, I just passed that apartment." So, whoever's setting the fires is in the building.

The people that lived in the building were trying to get into the fire apartment, they thought somebody was in there. They were throwing water on the fire with pots and pans, but by the time we got down there a couple of rooms were going.

This was in the dead of winter. And, it is very disheartening to see young children and babies out in the hall. It was freezing cold, and the windchill factor was 13 below 0. There was ice in the hallway. They only had their stoves for heat.

I felt helpless. I wanted to help these people, but I couldn't. I knew that some fuck in the building was setting the fire. Later, we learned that the super of the building was the arsonist.

I got a medal for my part in the case. It was the Martin Scott Medal—it is named after a former chief fire marshal, and I also got a class 3 for apprehending these people.

There have been many changes since I came into the Bureau of Fire Investigation, like the development of the

•

training program for new marshals in 1980 by John Stick-evers.

The average guy that comes into the bureau has 17 years of experience as a fire fighter, which means he has existed in the monastic, parochial world of the firehouse all that while. As fire fighters they already have the knowledge of what fires do, but they need guidance on how to conduct investigations by asking the right questions, and putting down everything people tell them in an articulate form.

We also emphasize pistol proficiency. The cops shoot once a year; we have to qualify three times a year.

We now have the capability of easily gathering information that can help us put the arsonists away. We have computer terminals connected to motor vehicles, finance and the buildings department. We can also check names and criminal records with New York State Criminal Records and with the FBI.

We have had to fight to get all of these things, and now that we have them, we have to fight to hold onto them.

•

# 6

# CHIEF FIRE MARSHAL

## JOHN REGAN

### ( R E T . )

I came on the fire department in January 1959. In October of that year, I went to a second-alarm fire in a haberdashery. I was working in 46 Truck with Hughie Gilroy. As I look back at him now, Hughie was about my size, 52 years old, a nice man.

The rig was an old four-wheel drive. It had a running board where you could stand. The running board was about 24 inches off the deck.

When the alarm came in, Hughie slid the pole, but his pants came down around his ankles when he reached the apparatus floor. He shuffled over to the truck and tried to light his foot up the 24 inches, but he couldn't because the trousers were restricting his movement. So, he kneels on the running board and the rig starts to move out. He has all he can do

•

to hold onto the rig. We go down the street with lights flashing, siren wailing and Hughie's ass sticking out.

We were the second-due truck. When we arrived 106 Truck had already entered the store. There was a good deal of fire rolling out the rear of the place.

In the front of the store was a good-size foyer where they had display windows. In that area they had a heavy hanging ceiling made of wire lathe and plaster suspended on iron rods.

As soon as we step into the foyer I looked up. And just like that, I saw the ceiling run a seam the whole length of the foyer. It opened up as if you would open up the pages of a book. I said, "Hughie, that ceiling's going to go."

He said, "Go back to the truck and get a 10-foot hook."

I turned and took two steps and as I hit the sidewalk, bang! The ceiling came down and seven guys were trapped under there. We went back in and tried to lift the ceiling to get the brothers out. We eventually got them all out.

I then went outside and sat on the rig. This officer comes along and asks, "What's the matter?"

"Oh God, I think I hurt my back."

He said, "You better get your ass up and get in there again. You're too young on the job to be talking about back injuries."

So, I got up and got my ass in there.

My back continued to bother me for many years afterwards. I finally transferred down to Ladder 171 in Rockaway. I thought 171 would be a lot quieter and I could hide out with my bad back. I was only down there a month or two when one afternoon I just couldn't walk anymore. They took me to the hospital. I went into a chest cast for six weeks. It was really an experience. That's when you find out if your wife really loves you or not.

When I got back to work they put me on the limited service squad. I wound up in First Deputy Commissioner George Mand's office. He was a very stately-looking man. My job was to sharpen his pencils and make sure his carafe

•

was full of cold water. I was in my early thirties and I thought to myself, "Oh, how do I do this for 20 years? I have to get out of here somehow."

Then I noticed these guys walking around with suits on. "Who are these people?"

"They're marshals."

"Well what do they do?"

"Nobody really knows a great deal about what they do."

I swear this was the attitude then. I said, "Well how do you get to be a marshal?"

"Oh, you've got to be a former cop or be a former insurance person or have political weight."

My kid brother was working for a first-term congressman by the name of Hugh Carey. I said, "Tom, how about getting a note from Carey to the fire commissioner telling him what a marvelous guy I am."

Tom said, "Yeah, sure, no problem." He's my brother, what else is he going to say.

I got called down for an interview and they said, "OK."

My first job as a marshal was on Lincoln Place just off Washington Avenue, in the 80th precinct in Brooklyn. The fire was in the closet of a second floor, left apartment of a four story, brick tenement with two apartments on each floor.

I'm by myself. In those days you worked alone unless you were doing the night tour. And you had to get experience doing routine fires before you were allowed on the night list. Also at that time the battalions didn't report fires as "suspicious." Instead we had a daily printout from each borough and Barney Riley, the deputy chief in charge of Brooklyn, Queens and Staten Island, would assign the routine work. A fire in the cellar, a fire in the hallway, a fire in the closet, a fire in the bed was always investigated.

So as a matter of routine I picked up this fire. When I got out there, I talked with a young girl. I asked her, "Do you know the people who live in this apartment?"

"Yeah, I do."

•

"Well did you see any of them here today?"

"Yeah I saw Mr. Blank* just before the fire. I was just coming down from lunch and going out to play. He came up the stairs and he had keys in his hand."

I said, "What keys?"

"Keys to the door."

This was just conversation, it didn't mean anything to me at the time.

I get a hold of Mr. Blank and start the interview process. He tells me, "Yes, my girlfriend and I are having a spat, but I couldn't have made the fire. . . . I don't have a key to the apartment."

Now who are you going to believe, him or this little 10-year-old girl? So, I asked him, "Where were you at the time of the fire?"

"I was across the street in the gin mill."

He then said that he got in through the first door, because it was left slightly ajar, which is possible.

Even though that little girl didn't see him go into the apartment, I thought to myself, this son of a bitch has a key.

He reaches into his pocket and says, "Here are my keys."

I tried the keys in the front door, but they didn't work. Then I thought, "This son of a bitch hid the key." If I were him, sitting in that gin mill across the street, I'd take a piece of gum and stick it under the bar. So, I went back into the tavern, and I run my hand all along under the bar. There are wads of gum all over the place, but no key. Now, my hand is dirty, so I walk into the men's room to wash it off, and after doing so I dry it on this cloth towel. Then for some reason I pull the top of the towel dispenser slightly away from the wall. Bang! Into the basin drops a key.

I take the key, go across the street, and it fits the apartment door. I said to him, "Come here. Let me show you something. You took this key. . . ." And he gives it up.

•

In retrospect I felt like Columbo, because it was dumb. I now realize that finding that key didn't mean anything, but at the time, for me the key to solving the crime was the key.

I haven't thought about this in over 30 years, but there was this fuck out in Queens. This guy was not a nice man. He took advantage of some elderly people. He was a black man and the people he took advantage of were black.

I can remember talking to an elderly man. I asked him, "Why did you call the fire department?"

"Well, because I saw a fire across the street."

"What else did you see?"

He said, "I saw something else, but I really can't tell you, because I'm terrified of this man."

"Why are you terrified?"

"Because he's an ex-convict, and he is a killer. He kills people. Mister, let me tell you how scared I am. If that man came into this house and wanted my wife, I would give him my wife." There was real fear in his heart.

Well, I spent a great deal of time talking to him. Finally he told me he saw this bad individual exit the building immediately after a fire. He watched him from his front window. The fire was across the street in a private dwelling that this individual had turned into a rooming house, though he didn't live in it. The elderly man also gave me enough information on this individual so that I could locate him, because he did not actually own any property. He took it over: "This is mine." Then he would move people in, and he collected the rents. I mean this guy was a tough piece of work.

I didn't want to take the case on by myself. I considered myself much too young in the job. So, I reached out for an older guy. I got hold of Ernie Bower. His brother is still on the job. Ernie has since died. But Ernie was a very active marshal in Brooklyn, and he was a good guy. He was well-thought-of by all the guys in the office.

A young girl who's renting a room on the first floor in

•

another house tells me that this bad guy is coming to collect the rent. Bower and I go into her apartment and wait. We spend half a day waiting, and, of course, just as we're ready to give up, the iron front gate opens, and there he is. He comes walking down the entranceway, and knocks on the front door.

Now I get against one wall, and Bower gets against another wall. We've agreed to let her open the front door. She ushers him in. He is only two or three steps inside the front door when I grab him, and throw him up against the wall. Bower is right next to me. We're shouting at him, "Spread eagle. Spread eagle."

He says, "Okay. Okay. You got me. . . . Who are you guys, the FBI?"

"FBI? FBI? No, BFI. We're fire marshals."

It was the funniest thing at the time. You talk about a man who had just reached the depths of life. He was so pissed off that fire marshals got him. He just could not imagine it.

We go over to the police station. Ernie Bower leaves, but I'm sitting in the back room taking a statement from this guy. There's only the two of us.

As he had intimidated the elderly man and his wife, he tried to do the same with me. He asked, "You have a family? Where do you live? You live in Brooklyn, I think. Don't you? And you have a wife?"

I don't consider myself a violent individual. Today, I don't even carry a gun—though it's not the gun that makes an individual violent. But I went over the desk after him, with my gun out. I stick it down into his mouth. I can hear him choking on it. I yell, "You motherfucker. I'll tell you—you even have that thought again in your fucking life and I will kill you—cold dead." It scared me. "Be bad with me. Kick my ass, whatever you have to do, but do something to my children, to my wife? I don't think so."

After this little to-do, we reached an understanding.

•

Later, I testified in court that this man made statements in an attempt to intimidate me and that we did have a discussion. I wasn't going to let him get a leg up on me in court and say that I threatened him.

I'm by myself again. I have to check out the aftermath of a small fire in the rear, of the second floor, of a four-story brownstone. This type of building has a front stoop. So, I walk up the front steps, and bang on the door, and out comes this little black woman, on crutches. She is whacked out of her head. She is stoned. She only has one leg, and she's having a difficult time balancing herself.

I said to her, "I'd like to get in to see the apartment where the fire was."

She says, "You ain't going in there."

"Oh, lady please, I'm from the New York City fire department. I'm a fire marshal," and I present the badge. "I'm going in there, it's just a matter of when am I going in there." It is sometime in the afternoon, and it is a very, very hot day. "Please let me get in, and get a look at the rooms, so I can get out of here. I don't want to interrupt your life. I don't want you to interrupt my life. Please, just let me go. . . ."

"I'm telling you that you ain't going noplace. You ain't going in there."

"Oh God," I thought, "it's 90 degrees, and here I am in Brooklyn arguing with a little, drunken lady with one leg. How the hell can this be? This is not right."

I reach in my pocket and take out a pack of cigarettes. I take one out and put it in my mouth. I close my eyes, and say, "Lady, I'm going in there." I got the cigarette dangling from my lips, and I'm reaching into my pocket for the lighter. With that, whack! She gives me a shot in the face, and knocks the cigarette out of my mouth, and

•

almost knocks me down the stairs. I just lost my balance for a second.

I thought to myself, "Oh God, look at this, now you're going to lock this woman up. Nah, I'm not going to lock her up. It's 90 degrees and it's August, and I have a place down on Breezy Point, and I'm going to go and have a nice swim. I'm going to come back tomorrow and get into this building." With that, I walked away. I just could not picture dragging this crippled, drunken woman into a courtroom, and saying, "she hit me." They would tell me to get the hell out of there with that story.

You have to understand that there are different reactions to violence. You can stick the muzzle of a gun down a guy's throat until he starts gagging, and then say, "I can't hit this lady." Temper levels are so strange and so different. I guess that's one way you can understand how every once in a while a mistake happens, and a young cop shoots when he shouldn't. It's really tough to be up all the time, and to be right all the time.


Sonny Callendro is my partner. Sonny is a very light skinned black man. He is a very nice, very quiet, very religious kind of guy. I'm breaking Sonny in.

One night we go to this apartment, in the East New York section of Brooklyn, where a Molotov cocktail had been thrown through the front window. A young black girl is the occupant of the apartment. We talk to her for a little while. I said to her, "Do you have any idea who did it?"

She said, "Yeah, I know exactly who did it."

"Who's that?"

"My boyfriend."

"Well, how do you know that?"

"Two of my cousins saw him doing it."

I said, "Oh."

We are standing at the front window when she says,

•

"You see the fellow standing down there near that pickup truck?"

"Yes."

"That's him." There is a crowd of maybe 20 or 25 people standing in front of the building, and he's just slightly to the left of them.

I said, "All right, I'm going to go down, and take him over to the police station for questioning; but I want you to contact your relative who saw him, and send him over to the precinct. Okay?"

I start to walk away when she says, "Wait a minute. I have to tell you something."

I said, "Yeah?"

"He's got a gun."

"Oh, thank you very much." It's a very comforting thought to at least know what you're up against.

Sonny Callendro is standing beside me. It's a little cold, and we both have on topcoats. I said to him, "Please do me a favor, leave your gun in your pocket, and stay to the right of me." I didn't want Sonny getting excited with a gun to my back. Those goddamn guns scare the hell out of me.

We get down to the street, and I have my gun in my hand, but I have it concealed.

I point to the boyfriend and say, "You, stand to the side. You see this?" As I'm talking to him I'm raising my palm. I didn't want to go out and start pointing into a crowd of people with a bloody gun. But I'm just showing him enough of the gun for him to know that I have the advantage. I then say, "Just stand right where you are, and put your hands on the truck window." I'm walking towards him all the while, when I hear, *Bang*! A sound that I wanted to hear. It is the sound of his gun dropping into the cab of the truck. He had it in his hand.

Thank God, I had a few minutes' conversation with that girl, and by being decent to her she was decent to me. She said, look out. Sometimes that's all you need.

•

*     *     *

The most fascinating case I had was the Puerto Rican Social Club fire which occurred shortly after 2:30 in the morning on Sunday, October 24, 1975. The club was on Morris Avenue and East 165th Street in the South Bronx. Twenty-five people were dead on arrival and another 24 were seriously injured. Two of the injured would later die.

The club was a 25- by 50-foot room on the second floor. There was a single stairway leading up it. In the front left corner was the bandstand, behind which was a closed running steel door which leads to the fire escape. Along the rest of the front wall was a series of windows. In the right rear corner were the bathrooms.

I'm sure it had to have been horrible in there: The panic of trying to get out, because of the fire coming up that stairway, through the only means of escape, and just pouring into that large open room.

Those people that survived managed to do so by breaking the front windows and jumping to the street. However, many of the bodies inside were found piled on top of one another near the broken windows, where they died of asphyxiation before they could reach the opening.

I guess people, adults react pretty much the same way as children do in a fire where if it is out of sight it is out of mind. So, they go under the bar, or into the men's room, or into the ladies' room to avoid the fire. Those were the places where the rest of the bodies were found.

A police sergeant on patrol saw the flames and turned in the alarm. The fire department had most of the fire knocked down within 5 minutes and had it under control in about 15.

We determine that flammable liquids thrown on the stairs and in the entranceway produced this fire.

The fascinating part was just trying to reconstruct what had gone on. At the time I knew everybody in the club. I

•

knew who was at each table, and this was from interviews and reinterviews.

I've been to a hundred of these dances. Social club is just a name—I mean these types of events go on in church basements all over the city. People just out for a little weekend's enjoyment, and to me it is very, very sad, because I really could relate to it. I've been there.

All the guys in the band had two jobs, and many of the survivors were also working a second job, or mama and papa both worked.

Mario Merola, who was the Bronx district attorney, the Bureau of Fire Investigation, and the NYPD formed a task force to approach this fire. At that time it was the largest case of mass murder in New York City.

We worked out of the 48th precinct. Tommy Flannagan was actually in charge of our side of the investigation, but later Tom goes on vacation. Mike O'Connor, who was then the chief fire marshal, asked me to take over the investigation. He said, "If the investigation is solved while you're in charge of the fire end of it, I'll make you a supervisor." I didn't think about a reward, because the case was so fascinating, and I was too involved with the intensity of what was going on.

At first, the police department believed that the social club owner, who was in the club at the time, set the fire. He had some sort of insurance policy. He had gone out for orange juice, though he came back into the club. However, in my experience a man is not going to make a fire, and then walk into the building, and trap himself upstairs, unless he's suicidal. It's definitely a first if it happened that way.

We worked out an agreement. The police were going to center their investigation around all of the people in the soical club, at the time of the fire, and I had everybody else outside the club. This opens up all of New York City to me. I thought, "Holy Jesus Christ, these people scare

•

me, because they don't even know what they gave me."

One of the people we were in contact with is a fellow by the name of Stanford Beilue. His family is originally from New Orleans. Stan was a very reluctant witness. Stan fascinated me. He walked on the edge, but I thought he was a good man.

Stan was driving home the night of the fire. He was driving up Morris Avenue, and he stopped for a red light. The social club was like three-quarters of the way up on the left hand side of the street from where he's stopped. There's a car double-parked in front and slightly forward of the social club. Stan describes it as a green Pontiac, and he gives it a year, because he had a similar vehicle. As the fire erupts that car quickly comes towards him, with no lights on, makes a right-hand turn and is gone.

The reason that I want to talk to Stan is that New York license plates were iridescent, and if his lights hit that plate maybe he remembers the number. Unfortunately, Stan did not remember the number, but he eventually agrees to be hypnotized. We are hoping it will facilitate his recall.

The hypnotist is brought in and Stan is put in a trance. Then the hypnotist briefly talked to him again. Everyone in the room is quiet. In a little while Stan says, "It's raining." Never before in any of our conversation did it ever come up that it was raining the night of the fire. "It's raining, and there's water on the windshield." We later found out there was a slight drizzle, but it was there and gone.

He was seeing something. He again describes the green Pontiac. He does see the plate, but he is still unable to tell us the numbers on it. He does tell us that he sees three young men getting into the car.

The investigation is going on for a long time: Eight, or nine weeks, or longer. Over 1000 interviews were conducted, but this case was solved the same way that most crimes are. There's a guy caught on the Cross Bronx Expressway in a stolen car, and he wants to get off. So, he

•

says to the ADA, "Listen, if I give you somebody involved in that social club fire, can we make a deal?" The deal is made and he gives up the name.


But the information came from the Bronx DA to the police department. I was there that night—it could have been Christmas eve. Our office was on one side of the hallway and the cops were on the other. Donald Washington comes to me and he says, "Something's going on over there."

I asked, "Why do you say that?"

"Because there's a lot of scurrying around over there."

I go across the hall and ask, "What's up fellows?"

"We're just going to go out."

"I'll go out with you."

"Nah, you look too much like a cop,"

"I said 'well, allright,' we'll send Jack Lovet." Jack was a big Irish-looking guy.

"No, no, he looks like a cop too, and we just want to go out and pick somebody up, and bring him in."

"Well, take Donald Washington. What are you going to say, fellows? Give me a break." Donald is a small, slight black man. "Does he look like a cop too?"

So, we get two cars and we all go out looking for a 17 year old by the name of Hector Lopez. We pick him up, and he says, "Yeah, we did it, but we did it because José Antonio Cordero asked us to do it. We were drinking rum with him for a couple of hours before, and we did it as a favor to him."

Then we get 40-year-old José Antonio Cordero. The 7th homicide detectives arrest both of them, and hustle them right over to the precinct.

They have Cordero upstairs, in a room with a table where 8 to 10 people could sit down and take notes.

I'm now in the room alone with Cordero. Suddenly it dawns on me, "Holy shit, look at this, Regan is in the

•

room with a guy that just killed 27 people," and nobody has searched him to my satisfaction. I tell him, "Get over here against the wall," and now I start going through his collar, because they used to keep razor blades there. I go down both of his legs to his ankles and come up under his groin and take a handful of balls, because I ain't giving him too many shots at me. If it's going to be unmasculine, don't worry about it, I'll take a shot.

Well, Louie Hernandez, a detective there from 7th homicide, comes into the room after I just finish searching Cordero. I tell Cordero to sit down at the far end of the table and I sit.

Louie says, "I've just fucking well about had it." With that he laid his open hand on the table. *Bam!* The whole table vibrates, and Cordero almost shits himself. He didn't quite know what else was coming. I am sure he thought that this was the end of his world. I wasn't expecting it either, and I jumped just from reflex action.

On January 2, José Cordero and Hector Lopez were arraigned in Bronx Criminal Court for arson and murder.

Tom Flannagan comes back from vacation. Then Tom, Louie Hernandez, and I go to Puerto Rico to search for a third accomplice, a 17 year old, who they call "Beans." His real name is Francisco Mendez. We thought that his family was hiding him down there.

We looked for him for several days. With the help fo 30 local policemen, we finally found Francisco in the town of Arecibo.

Francisco was arrainged in San Juan as a fugitive, but he waived extradition, and we brought him back to New York on January 12. He was actually my prisoner. I had him cuffed when we got off the plane.

It was a night of revenge for Cordero. He was going to balance all of the books. There were two other smaller fires that he made earlier that night, but they never prosecuted him for them. I think that they didn't want to cloud the central issue of the social club fire.

•

Cordero was married and had two children, but his motive for the social club arson was his jealousy over a young woman who attended the dance that night against his wishes. He got his pound of flesh.

Cordero was interviewed two or three times prior to being arrested. He owned the same-year Pontiac we were looking for, but his car was blue. After his arrest we brought his car back to the scene, and put it under the streetlight. It looks green, because of the amber light, and that is what Stanford saw.

On June 16, 1977 Cordero pleaded guilty to murder by arson. However, his defense attorney, at a pretrial hearing on December 1977, said that José was too sick to have made a rational confession.

Later in court, when Cordero gets on the stand and testifies as to what transpired the day he was arrested, the question comes up, "Were you ever physically abused, or did anyone beat you, smack you, hit you, or forced this confession out of you?" He said, "There was one little fat detective that kept twisting my balls." And he looked all over for this little fat detective that was twisting his balls, but he never looked at the fire marshals. Believe me I was not twisting his balls. I was trying to safeguard my own testicles.

José Antonio Cordero was sentenced to life in prison. Hector Lopez, who admitted lighting the fire with a match, also received a life sentence. Francisco Mendez got 25 years to life for spreading the gasoline.

I left the bureau in February of 1980. I was going with the Department of Labor. I had a job as an investigator, grade 15, with the federal government. It was a nice job. You got a car, $27,000, and you worked on your own. After putting 17 years in the marshals' office, I just thought it would be interesting. I also had 4 years in the Marine Corps, and I thought I would put in 16 years with the

•

federal government, and I would get a second pension.

In October of that year, Joe Hynes, a neighbor and friend of mine, calls me up. I was sitting up in my bedroom on the second floor when I got this call. After our conversation I lay back on the bed.

My wife comes home from work, and there are tears in my eyes. She says, "What's the matter?"

I said, "I just got a call from Joe Hynes."

"Yeah?"

"He's going to be named the fire commissioner."

She said, "Oh, that's terrific. So what are you crying about?"

"I'm not crying, I just have tears in my eyes. He wants me to be the chief fire marshal."

"Doesn't every fire marshal want to be the chief fire marshal?"

I said, "I'm not sure."

"When he calls tomorrow you tell him that you'll take the job."

She certainly didn't have to force me into it, because I love the department. So, when Hynes called the next day, I said, "Yes I would be happy to take the job," and I thanked him for having this faith and trust in me.

Wednesday of the following week I get a telegram from the federal government saying the position has been eliminated.

It all worked out, thanks be to God.

My most difficult challenge as the chief was trying to convince the department and the City of New York that we offer an extremely important and valuable service, and what we do is very honest and very necessary.

When you go to other cities and they tell you that they don't have an arson problem, 99 percent of the time it means only one thing: Fire investigation is not being done. It only becomes a problem when you investigate it; but if you don't address it, it doesn't go away, it just gets worse.

Speaking of getting worse, we had a fire not too long

•

ago in another social club in the Bronx. The place was called the Happy Land Social Club, on Southern Boulevard.

The local engine company was eight blocks away, on their way back to the firehouse. It was the quickest response that you could get—bang. They were there, and they had the fire knocked down in no time.

I was at home, and the first call I got was from the dispatcher informing me that there were four DOAs. I was up and starting to get my head together, because I was going. Four dead is serious. I mean as serious as you can get. But while I was still on the phone with him, he said, "Now they're saying they have 20 DOAs."

I said, "Oh, my God," and I start getting dressed. This was at 3:41 in the morning. Then I called Tommy Clarke. I said, "Jesus, Tom, we got to get to the Bronx. I need you. I need some help. We may have some serious problems up there."

He said, "I'll be ready in a minute."

I got out as quickly as I could, and I picked Tom up, and we went directly up to the Bronx. We were there in short order, certainly before five, I guess.

When we approached the building from the sidewalk there were no second-floor windows to indicate there was in fact a second story. There are two doors into this place on the first floor. The building is about 20 by 60 feet.

The fire occurred on the first floor, just inside the entrance on the left. Someone who immediately became aware of the fire runs through it and exits the building. He was taken to the burn center at New York Hospital. A woman in the cloakroom unlocks the door on the right and exits unharmed with three other people. The remaining 87 people in the club died. The majority of them were in the main room on the second floor.

When we went up to the second floor the bodies were scattered all over the place. Some were still sitting at tables, others were piled on top of each other.

•

The disturbing fact to me was to see the ages of these people. As I get closer to 60 I'm more accepting of death, and the eventuality that it will come to me, but it just seems so terribly unfair that these young people were dead. It is overpowering to understand, or justify how quickly the lives were taken from these 87 people. A human being is so fragile.

Like the Puerto Rican Social Club fire, this one was also brought about by jealousy.

Dealing with these tragedies doesn't get easier. Maybe we do get softer with age.

My career with the fire department has been a great experience. I have been fortunate enough to be in a job where I honestly believe that I have contributed. I have a great deal of personal pride in the marshals' office, and in the fellows that continually perform the work. I have been here just to guide them a little bit, but I can't tell you how rewarding it is.

•

•

# 7

# DEPUTY CHIEF

# FIRE MARSHAL

# TOM CLARKE

had been fighting fires for 10 years in the Red Hook
section of Brooklyn. Then in August 1977 I get a phone
call from the chief fire marshal asking me if I wanted to
become a provisional fire marshal. They were rapidly ex-
panding the BFI, and they were looking for guys who had
been cops or had college degrees. I had graduated from
college the year before, and I put three years in the police
department before becoming a fire fighter. I turned him
down.

That night I went to work in Ladder 131. I told my
lieutenant, Hugie Steen, who I really respected, about how
I turned down the offer to become a marshal.          He
said to me, "What? Are you more stupid than
you look? This is the opportunity of a lifetime to
see a different aspect of the job."

•

Several days later, I called up the chief's office and asked to be reconsidered. Three weeks later I was in the next fire marshal class.

In 1979, we were working for John Regan. We got called to Staten Island for a 10-45, code-two; a child was seriously burned. We determined it was an accidental fire. If it wasn't for a black guy who went by the name of "Corn-row," she would have been dead.

She was recently awarded five million dollars in damages. They sued six companies who settled right after the examination before trial. Her attorney showed pictures of her. She was the most disfigured person I have ever seen.

While we were doing our initial investigation we got another call for an incendiary fire in SoHo. This was a rapidly gentrifying area in lower Manhattan.

We finished up at the hospital, and responded to SoHo. We got there an hour and a half after we received the radio call. The manpower was really stretched thin, just like it's getting today. When we arrived there was only one engine, a truck, and the battalion chief still on the scene.

The chief began screaming at us, "Where the hell were you? What took you so goddamn long? We almost lost 35 fighters tonight."

I could understand why he was so hot under the collar. They were on the scene for some time before they found the fire in this kosher meat market. Above the meat market was a 12-story, fully occupied residential building. The individuals who set this fire had put flammable liquids in containers with electrical timers and heaters to ignite it in two separate locations in the store. These guys, however, were amateurs, because only one location went off. But when the fire fighters moved in to fight the fire in the mezzanine area they had to pass by the area with the as-yet unexploded gasoline.

We start doing our investigation. In a wastepaper basket we find the boxes that the electric timers came in. On top of the boxes are remanents of sandwiches, and numerous

•

coffee cups. I knew this was going to be good circumstantial evidence if the defense of the owners was that their place was burglarized and then set on fire. I mean, what burglar comes in with timers and is such a neat guy that he puts the boxes in the wastebasket? Oh, and he stays there to have sandwiches and numerous cups of coffee in a building he intends to burn down.

In questioning people in the neighborhood we found out that the two brothers who owned the store had been in business several years, but the day before the fire one of them gave his card to the superintendent of the building across the street. He said to the super, "If anything ever happens, if a pipe breaks or something, I live in Monsey, New York. You can get in touch with me."

Monsey is a town upstate where many Hasidic Jews live. These guy were associated with them, but they didn't have the curls or dressed in black.

One of the brothers was on the scene, and one of our questions to him was, "How did you become aware of the fire? Who notified you?"

He told us he was notified by the police department. Later, we subpoenaed the phone company for his records. He got a call that night from the super of the building across the street at 1:15 and he gets down there 40 minutes later. Which he just couldn't do, unless he was driving 90 miles an hour and not stopping for lights.

There was a little altercation that night, because he wanted to come into the store, but we had declared it a crime scene. Of course, what he wanted to do was disturb the evidence. We had a cop on the scene, and we asked him to keep this guy out of the building and out of our way.

I'm inside the store, when I hear a disturbance going on. I run outside and ask the cop, "What is he doing?"

It seems he jumped up on the fire escape and pulled this louver out. "See, this is how they got in!" he said.

•

We found out that he had made numerous complaints to the police that he had been burglarized.

I used to work on the side, installing windows, and I determined that it was not the original mortar around the louver, but new mortar. So, we concluded that his intention was that the fire fighters would knock it out when they saw the smoke pouring out. However, it was by the secondary area which didn't catch on fire. He tried to say that the burglar just placed it back in there.

We concluded that it was a key job—that is, one of the four key holders opened the store up and set the fire—but only the brothers had anything to gain from a fire.

We had customers who told us that these guys had many pictures of their family and the old country in beautiful frames on the walls, but despite the soot we could see that they were missing when the fire started. When we asked him if he took the pictures out he denied it.

Later, when we wanted to question both brothers we asked them to meet us at the precinct. The brothers went in their cars; we went in ours. We had the door open when we questioned them. This way they could not say that we had them under custodial arrest. We questioned the brothers separately, but they never gave it up that night.

We had probable cause to arrest them, but we wanted to wait and do it the right way because a summary arrest closes off many avenues of investigation, and a good lawyer can get the case thrown out. It took 11 months to conclude this investigation.

There was some reluctance on the part of the Manhattan DA's office to take this case, because it was largely circumstantial. This is where Brian O'Donough, the ADA, comes in. We brought him to lunch at the firehouse, so he could meet some of the men who would have been killed if the unexploded gasoline had gone up while they were in the store. Each gallon of gasoline in the vaporized state is equal to 16 sticks of dynamite. The two brothers set out five gallons of gas. O'Donough took the case.

•

We checked their financial records. While they looked good on the outside, we found that every year they were in business they were losing money.

Before we went to trial there was a hearing to determine the admissability of certain evidence. As an ex-cop I vouchered most of the physical evidence we collected with the police property office, but I kept the wastepaper basket, with the food residue and the coffee cups still in it, in my locker. I knew that if I turned it into the property office they would just dump everything as if it were garbage. We received permission to introduce the basket and its contents as evidence in the trial.

Their defense attorney cost them $80,0000. They tried to get us on the gasoline. The defense asked us, if we vouchered only a few milliliters of the gasoline, where was all the rest? We told them that our boss, John Regan, told us to empty the containers of it into the department cars.

The jury, thank God, was composed of reasonable people. They didn't fall for the idea that some burglar crawled through a small vent with five gallons of gasoline, sets up the timers and the heaters, and puts the gas in buckets that were indigenous to the store, then has a meal with coffee, puts his garbage in the wastepaper basket, and leaves with the gas can. They were both sentenced to 5 to 15 years in jail, and the convictions were upheld upon appeal. While out waiting for the appeal they both fled back to Eastern Europe. One brother subsequently returned to New York and served his time and then moved out west. The other brother is still a fugitive.

This was one of my most rewarding cases, and I learned a lot doing it. Hopefully you're always learning something new. It renewed my interest in the marshals, because until then every year I was going back to my first love—the firehouse. But I knew I could be more helpful to my brother fire fighters as a marshal.

My most interesting case, however, was one where I worked with the FBI in 1980.

•

An FBI agent—Roger Viadero—comes to John Regan, who is now the chief fire marshal, and tells him that one of the U.S. attorneys is looking for someone knowledgeable about fire insurance.

So. John Regan, Bernie Casey, and I go over to the office of the U.S. attorney for the Eastern District, at Foley Square in Manhattan, and we speak to Sarah Gold. I liked her; she was a real cocky lady.

She tells us that there is an organized-crime arson ring that they want to infiltrate. One of the principals in the ring is willing to wear a wire, because they have him on another charge, and he is facing some heavy federal time.

Gold goes on to say what the job will be of the guy they are looking for. It is a dangerous undertaking. All of a sudden I understand why I was invited to this meeting: Everyone's eyes are looking at me.

We go back to our headquarters at 110 Church Street, and John goes into and speaks with Commissioner Hynes. I then go in, and the question is posed, "Do we have anyone willing to undertake this assignment?"

I said, "Yeah, I'll do it."

At our first meeting we learn that one of their torches is now out in Las Vegas. It is decided that the FBI will pick him up and place him in custodial arrest, and maybe we can get him to obtain information in addition to that which will get picked up on the Nagra.

Two agents from the Las Vegas FBI office go to where this guy is living. But they are informed that he died two days earlier in a pool accident.

"Well, how old was he?"

"Oh, he was 34 years old."

However, there was no autopsy. Out there they don't have medical examiners; they have coroners. The FBI has his body exhumed. There was not one drop of extra water in his lungs. Somebody did a number on him, and this is what we are dealing with.

The FBI buys me some nice clothes to fit the role I'm

•

going to play. I'm supposed to come from plenty of money, and my father, brother, and I want to burn some of the many properties that my dad owns. My brother's part was going to be played by an FBI agent.

I'm Irish, but I am often taken as being Italian. I am married to an Italian sweetheart for over 20 years. So I took her maiden name as an alias, because I knew that I wouldn't forget it.

The first meeting we had with some members of this ring was horrendous. We were in a big restaurant on Long Island. There were two FBI agents in the restaurant, but I'm isolated from them, and there are three agents and the U.S. attorney on the outside.

I had a couple of drinks, and the adrenaline is flowing. I hand the main principle, who we were looking to get, an insurance policy. We had an insurance policy for $500,000 on some buildings on the Lower East Side. He questioned the policy. He said, "Where these buildings are is redlined. How can you get a policy from a legitimate company? It shoulda come from Fairplan."

The FBI had gotten the policy from the Globe Insurance Company. Initially the insurance company wanted the FBI to sign off on it, just in case there was a fire. The FBI said no, but they managed to get the policy anyway.

The situation was getting very tense. So, I jumped up and yelled to Paulie, who was the guy wearing the Nagra, "Paulie, let's get out of here. Fuck him. Didn't you tell him where you and I come from? All the deals we did over the years." Then I looked back at the guy and said, "If you don't think I can get a policy in this day and age, then you don't know who the fuck you're dealing with."

I was afraid they were going to find me out. There were three or four other wiseguys sitting at the table, and I thought I was in deep shit.

The main player goes, "Sit down! Sit down! Paulie, I like this guy. I couldn't get a policy like this, even with all my contacts." We talked some more.

•

Paulie and I start back to New York City, but on the way we rendezvous with the other FBI agents. I tell them I got into a bind, but all I get is blank looks. The transmission from the Nagra was not coming in that good, and I knew they didn't know that I was ever in trouble.

When I got back to headquarters I told them, "I would do anything for the job, but I need a backup who understands what's going on. I need my partner Jackie Brill." God forbid, if anything happened I knew Jackie would be there, not that these other agents wouldn't.

Anyway, Jack was assigned to the case.

The FBI gave me $500 to pay for dinner the next time Paulie and I were to meet with the members of the ring. I really got into the role-playing; I felt like a rich guy.

When the next meeting came about it was on payday, so I had several hundred dollars in my pocket in addition to the FBI's money. We had drinks in the bar before sitting down to eat. I gave the bartender a $50 tip. Well, the service in the restaurant could not have been any better. They couldn't do enough for us. When the bill came it was more than the 500 the feds gave me.

The FBI broke out in a sweat over getting the additional $150 that I personally had to lay out.

Paulie was a bad guy. He was wearing the wire, and he would say things like, "Tommy, aren't these lobsters delicious? Let's order a couple more." That was just to annoy the FBI agents who were listening, as they sat in their cars eating sandwiches.

We used some excuse not to burn the building on the Lower East Side, because there was no guarantee that people wouldn't get hurt if this building got torched. We went to Jersey City, where we had an isolated building. The FBI wanted to catch them actually starting a fire, but I pointed out to them the danger that still existed for the fire fighters.

We had meetings over three months, because I told these wiseguys that I wanted to start renovating the buildings before they burned them down.

•

Before these meets I always give up my gun and my identification in case I get tossed. We were late for a meet in Jersey City. We were trying to tape the Nagra to Paulie, while looking out for our contacts, and the traffic was tied up on Kennedy Boulevard, which leads into Jersey City. I forget to turn over the snub-nose .38 Chief's Special in my ankle holster. I didn't enjoy that meet, because I was afraid that were going to spot my gun, which in those days was the type carried by most law enforcement officers. There was a lot of time and money invested in the investigation, and I didn't want to blow it.

Some of these meets I enjoyed. I can see why some guys really get into role-playing and go undercover for long periods of time. When you are role-playing you're part of the scene.

I detested Paulie, but at times I found myself aligned with him. Paul drove a stolen Ferrari and packed a .25 automatic. The FBI let him do these things, but I never trusted him. He was a dirt bag.

My job was to generate conversation. I was not to do much talking, but to listen. I would ask *what-if* questions. These wiseguys were to implicate themselves.

This ring had collected on millions of dollars worth of policies. They didn't always collect the full value of the policy, because the insurance companies would bicker over the circumstances or the conditions of the buildings.

We had enough information to arrest their accountant. It was a rainy night. Jack Brill, two FBI agents, George Dwyer and Jim Roth, and I are standing on the steps of the federal courthouse on Foley Square in Manhattan. The accountant comes along. He is on his way to pick up a bogus law degree, when he runs into us.

Dwyer said to him, "Stop. FBI, you're under arrest."

The accountant looks at me, smiles, and asks, "Tommy, who are your friends?"

"FBI. You are under arrest."

"Tom, enough is enough."

•

Then one of the agents asks him, "Do you know Fire Marshal Tom . . ."

Well, his face just went blank.

The FBI picked up one of the torches on a Saturday morning in the Bronx, as he was exiting the church where he was just married. There were over 200 guests at the wedding.

I said to the FBI, "Aren't you being a little cruel?"

They said, "This way he can think about losing his new life with his bride, and maybe we can turn him to work for us."

Not too long after that, the FBI told me that I was not to go to any more meetings. I think there was a contract put out on my life. Maybe they found out that Paulie or somebody else was playing both sides of the fence. However, they did enable me to start packing an automatic, before the restriction on our guns was lifted.

Not long after I finished this investigation, I was sent to the FBI Academy in Quantico, Virginia for three months' training.

I like working with them. They have money. We have done other operations with the FBI, the DEA, ATF, and the NYPD Intelligence Unit, but because of their sensitive nature I can't talk about them.


One of my saddest cases occurred on the Lower East Side. We responded to a 10-45, code one. The fire was in the bedroom, and it was quickly put out, but this guy is behind the bed and he is horribly burned. He has plaster under his nails, which told us that he tried to claw his way through the wall in attempting to escape this fire.

I see that there is something unusual about the burn pattern. There is also a pot on the floor next to the bedroom, and water is running in the bathroom.

The guy's girlfriend is on the scene. She is very attractive. She is from Belize, and her English isn't too good.

•

However, she tells us that he fell asleep with a cigarette in his hand, and when she discovered the fire she tried to put it out.

I go back and look at the burn pattern, and it appears that the fire didn't start on the bed, but started under it and then burned up.

We questioned her some more, and she broke down and told us how this guy was abusing her for several years. He would beat her. He had his friends come in and have sex with her and assorted other things. She just couldn't take it any longer. So, while he was asleep she crumpled up newspaper and put it under and around the bed, then she lit it on fire. He woke up, and jumped out of bed, but he was trapped by the flames. When she heard him screaming, she attempted to put the fire out, but by this time it was too big.

I felt sorry for her, and I didn't want to lock her up, but I had to.


This other time we went to a fire in Brooklyn. This woman's bed was set on fire by her boyfriend. He believed that she was seeing another guy. While we are interviewing her, the boyfriend calls up and tells her that he will be right over. So, we get ready to take him.

When he comes into the apartment he yells, "Where is the motherfucker who's fooling around with my woman?"

We jump on him, and quickly put the cuffs on. He looks at us and says, "Who are you, police? FBI?"

"No, arson."

"Arsonic. I never ever used no arsenic in my life."

He was still shaking his head as we left for central booking.

•

# 8

## CHIEF FIRE MARSHAL

## JOHN STICKEVERS

\
had been a fire fighter for almost six years, but I left
and went to the police department. It's an unusual switch,
going from fire fighter to police officer. Usually it's the
other way around. Before I went to the police department
I worked in Brownsville. We were one of the busiest companies in the city. I started out in 231 Engine, and then
transferred over to 120 Truck. They had just broken five
thousand runs when I left, which was phenomenal for
that time. I enjoyed coming to work and fighting fires.

When I got to the police department I won the Bloom-
ingdale trophy. The trophy is a firearm that is given to the
individual who comes out with the highest overall average
in their police academy class. The average is based on the
three phases of training that you go through: academic,
physical and the use of firearms.

I liked police work, and I knew that the fire
marshals' office had something to do with police
work. I also found out that there was going to be
a differential in pay between fire marshals and po-

•

lice officers, so I made an application to come back to the fire department.

In one of my earlier cases we had an individual that I gave the moniker "the Fabulous Dabulis" to. His real name was Walter Dabulis. Walter worked in an ice cream parlor on Flatbush Avenue.

We had a series of fires that occured on Parkside Avenue near Flatbush Avenue. At that time, because of the setup of the office, no one really knew what the other guy was doing, and these fires went on without developing a pattern. However, there were neighbors on the block who sent in letters to the fire commissioner that finally got down to the bureau. Someone in the office then looked up all the fires on Parkside Avenue and saw that there was, in fact, a pattern of pyro activity going on.

A surveillance was set up in an attempt to catch this individual. Also pictures of known pyros were shown around the local bar and grills to see if any of these people had ever frequented any of these places.

The surveillance was in effect for about two or three nights when Tom Russo and I approached the then-chief fire marshal Vince Canty to ask if we could take part in it. We were given the assignment that evening, starting at 9:00 and supposedly finishing up about 2:00 in the morning.

We went out that night and looked the area over. Then we went to some more of the gin mills and showed pictures around, but it produced only negative results. We later went back and sat on the block, but we really had no idea which building to watch, because a number of them on the block had been targets for this individual.

I looked at the list of previous fires, and most of the fires had been made down in the basement with the exception of one building. And that building had fires on all of the floors, in the incinerator closets. It was my belief that the

•

person that we were seeking lived in that building, because he went above the first floor which is unusual for a pyro.

We watched that building and the others on the block. We left to get something to eat over on Empire Boulevard. When we came back I decided that we would check the building that we had targeted from top to bottom to see if we would discover anyone out in the hallways, or run into someone that might be able to supply us with some information, because the fires that took place in that building took place around this time—around 2:00 in the morning.

We went into the lobby of the building which had stairways on either side of the lobby going up to the top floor. Our plan was to split up, and go up the separate stairways to each succeeding floor for a period of one minute to listen and observe the area. When we got up to the top we would cross over and come down the opposite stairway.

While we were still in the lobby, the door to the building opened up and an individual walked in. He came directly over to me and said, "Does a Peterson live in the building?"

I had a feeling that this was our man. It was just a gut reaction. I told him, "I have no idea whether or not Peterson lives in the building." That put him on his guard, but to throw him off I told him, "I am waiting for a friend of mine who lives on the top floor. We are going fishing, but if you really want me to I will go up and get my friend, and see whether or not he knows if Peterson lives in the building."

He bought the story. He said, "No, never mind," and he walked over to the mailboxes and started looking at all of them.

I felt for certain that this was our man. He then turned around and walked out. We followed him. We observed him go down the block, cross over, walk a little on, and then go down into the basement alleyway of a large apartment house.

•

We tiptoed up to the edge of the alleyway. We could see him from our vantage point.

He walked to the rear of the alleyway, and tried the doors to the basement, but they were locked. He came back up to the front of the alleyway where some garbage cans were stored. He opened up the top of one of the garbage cans, and took matches out of his pocket. He struck a match and held it down towards the rubbish in the can.

Before it ignited, I yelled, "Police! Don't move, or I'll shoot." He froze. We took him out of the alleyway, placed him under arrest, read him his rights, and put handcuffs on him.

We took him around to the old 67th precinct. When we got there we had to wake up the night-watch detective. There was only one on duty. The detective let us into the squad room, where we searched and uncuffed the suspect, and sat him down in a chair.

I had gone through his pockets and pulled everything out of them. I found a wallet that was bulging. I looked through it and found an enormous number of pictures of girls, and guys and girls together. I didn't know anyone who carried around pictures of that many different people. A further search of his wallet produced advertisements for women's undergarments.

We searched the rest of his person, and in his jacket pocket I felt a bulge. I asked him, "What is in there?"

He said, "Nothing." I put my hand into his pocket and removed the item that turned out to be a sanitary napkin that was folded in half. When I unfolded it, I discovered that it was a used sanitary napkin. I asked him during the course of our interrogation where he got that napkin from. He said it was from his girlfriend. I asked what her name was. He said it really wasn't his girlfriend, it was a girl that he knew. I said, "A girl that you know? She had her period and she allowed you to take her sanitary napkin?"

"Well, it was just some hooker." He finally admitted

•

that he took it out of the trash in the ladies' room of the ice cream parlor that he worked in.

The interrogation of this individual lasted a couple of hours at the very least. He finally admitted that he was the person who was responsible for setting all of the fires that occurred on a two-block stretch of Parkside Avenue. He also admitted being involved in setting fires around the corner from the location where most of the fires were set. In all he admitted to being responsible for 22 fires.

He said that he had, on occasions, gone down to Parkside Avenue and set a fire in one building and then gone across the street into the basement of another building and set another fire. One was a multiple alarm and the other was an all-hands fire.

He said he used to go into the building that we had targeted, and he would open the incinerator closet up and inhale whatever smoke or vapors were in there, and that would get him excited. Then he would take the sanitary napkin out at the height of his excitement. He would smell it, set a fire, and then autoejaculate. These were small fires.

Now, around the corner from where he had set some of these fires, a homicide occurred. A Kings County Hospital intern's wife had gone down to the basement to wash clothes. When she was down there somebody raped her and strangled her. They had a palm print from that location. We had taken our prisoner to the wrong station house for the arrest. He should have been over in the Seven-one. When we got to the Seven-one there was a task force in operation for the homicide. They took his palm print, but it did not match the print that they had. So, he was never charged or implicated in the homicide.

He fit the classic profile of a pyromaniac. He had the domineering mother, and an absent nonexistent father, and some connection with fire in his background. He had never been with a woman in his life. All of his friends were males.

•

After his arraignment the criminal court judge sent Walter to Kings County Hospital for a psychiatric examination.


There was another case on Seventy-seventh Street, across the street from the Museum of Natural History, in the Park Plaza Hotel.

Someone from the fourth floor—we would later find out he was a pimp—called the desk and said that there was smoke up there. He also called the fire department.

The fire department came immediately and extinguished the one-room fire.

We were notified that there was a DOA on the fourth floor. The supervisor was Joe Bendis.

When we arrived, I left the car, went into the lobby and over to the elevator. The elevator was being operated by a black man. When I got on the elevator we had to wait there for a few minutes in order for the fire fighters to get on and off the elevator, and sort out who was going up and who was staying down. They were in the process of taking up their equipment, and whatever else they had to do in the mopping-up operation.

I decided that you have to start your investigation somewhere, so I started talking to the elevator operator. I asked him what his name was. He jumped back and said, "Calvin Jackson." That had no real significance to me at the time, other than I knew that in that area if you stop 10 people on the street 7 of them would have had some previous involvement with the law, and you're asking this question might make them feel a little bit edgy.

While we proceeded up to the fire floor, he told me that he was down in the lobby when someone yelled, "There's a fire!" He quickly went up to the fourth floor where he saw smoke coming out of this particular room. He then assisted in stretching the line, and trying to put the fire out. That was it. I never got back to Calvin.

•

I entered the room. The fire was out. It was a very small fire, confined to the bed, and the person who was dead was in the bed.

It was one of the few times when the field forces had left the body exactly where it was when they discovered it; the way it's supposed to be done at a fatal fire. Usually I find that the members of the fire department love to save dead bodies. There may be a subconscious thing operating when they do this. If they can take this person out of the environment, then in some way they may be able to cheat death. Because I've seen them remove obviously dead people with their heads split open and the brains spilling out. I've seen them scooping these people up and putting them in body bags and getting them out with absolutely no reason to do so.

I had a first-hand view of the deceased: A white female, by the name of Winifred Miller, who appeared to be in her late forties. She was naked and lying facedown. She had two socks on. They had suffered some damage as a result of the fire. Her head was turned to the right and appeared to be pushed into the pillow.

Her hands were down by her side with the palms turned upward and the fingers curled inward. Her fingers were burned. Some of them were burned very severely, especially on the outer tips. So, the curling of the fingers may have been a result of the heat application. I believe that the curling was partially due to the heat, and partially as a defensive move in reaching back to try to grab her assailant.

She appeared cyanotic. As a result, I felt that she had been suffocated or strangled.

With the burning on her back I tried to ascertain whether or not she had any clothing on when the fire started, because it appeared to be a wispy type of burning. There were some pieces of burned cloth on her back. It turned out that this was the result of the hose stream blowing portions of the bed clothing up on top of the body.

•

She had blistering and marginal reddening of her legs. There was some deep charring on the legs, but it was the portion of the leg that was actually abutting the bed post. There was some burning to her genital area, and her hair was burned.

The fire in the bed appeared to be a very fast type. I felt it was as a result of the application of some type of a flammable liquid. It turned out that a flammable was poured over her and over the bedding and set on fire. The burning to her back resulted from a pooling of the flammable liquid in the small of the back. And the wispy burning was as a result of the flames coming off that pool.

I was teamed up with a detective from 4th homicide, and we started a concentrated investigation on Miller's case. We investigated it for a period of about three months.

We were trying to trace Winifred's movements, and find all of her known associates and companions, to see who it was who saw her last, and who may have had a grudge against her. Also we concentrated on people within the hotel who may have engaged in this type of activity before. There were a whole bunch of people in there that were suspects.

We got lead off Calvin Jackson early on in the investigation, because the hotel clerk had told us that this other guy, who reported the fire, had been down in the lobby 10 minutes before. And that fitted the profile of somebody reporting the crime that he himself committed.

Now this man who reported the fire wasn't a tenant, he was just living with a girl in her room. She was on welfare, and he was just a live-in lover. The backgrounds of both of those individuals indicated that they had a potential for committing this particular act. They had both been arrested for drugs, and she was arrested for prostitution. He was a totally unsavory-type character.

He did cooperate on the initial stages of the investigation. His cooperation was based on the fact that he was really not guilty, and he knew that he was not going to

•

get into trouble. However, after a while he became very defensive and refused to cooperate. Whenever he refused, there was some pressure brought to bear upon him. Whatever cooperation we later got was not given freely. He would cooperate only to avoid getting himself locked up for some other reason. He was a pimp. So we put some pressure on him by threatening to lock up all of his girls down in the Forty-second Street, Times Square area.

In our investigation we learned that Miller was a lesbian, and in tracking down her whereabouts prior to her demise we got to talk to many different people. I found out that there were places operating in the city that I didn't even know existed before. There were gay bars that catered only to lesbians. I found gay bars that catered to a mixed bag of lesbians and male homosexuals. Then there were some male-only homosexual bars. And people who worked in one place also worked in the other locations. It was a very interesting revelation to me for the period of time that I engaged in the investigation.

However, the investigation ground to a halt. We weren't really getting anywhere on deciding who was a prime suspect. There were many potentials, but the demands from my bureau were such that they could no longer allow me to concentrate on just this one investigation. So, I was pulled off the case and brought back to conduct fire investigations in the bureau.

A couple of weeks later Calvin was observed on a rear fire escape of an apartment house six doors away from the hotel. He was thought to be a burglar. The people who made the observation called the police and waited for them to arrive. They said that they watched him come down the fire escape and go into the hotel.

The police came into the hotel, and he was identified as the person who was the burglar. He was taken into custody, and brought back to the scene of the burglary. The police forced entry to the apartment, and they found a 69-year-old widow by the name of Pauline Spanierman dead.

•

She was lying on the floor. She had been either suffocated, or choked, and there was some evidence of sexual molestation. Some of her possessions had also been removed from the apartment.

Calvin was brought over to the 20th precinct and taken up to the 4th homicide zone, where he was interrogated. Calvin began to talk about the Spanierman homicide. Then he talked about the homicide of Winifred Miller. He also talked about seven other homicides that he had committed in the confines of that hotel, one of which was committed after Miller's murder.

He said that he observed Miller coming into the hotel, and he immediately ran up the four flights of stairs, because he knew where she lived.

She was a little bit tipsy, because she had been out partying. When she got off the elevator she had to walk around the corner.

There was a little alcove right by her door, and he secreted himself in the alcove. When she opened the door he pushed her inside. Then he closed the door behind her.

He forced her to strip, and lie down on the bed. He attempted to rape her. While he was doing so, he put his arm around her neck and began to apply pressure.

She knew that she was now going to be a victim of not only a sexual assault, but of a homicide. I believe at that time she began to fight him off. But he choked her to death.

And then he took a bottle of perfume—it has a flammable alcohol base—applied it to the body and to the bed and set it on fire.

He searched the room for anything of value. He took a bottle of liquor and a radio.

He then went to the room that he was staying in with another girl that lived in the hotel—I believe that was room 225—and he deposited the liquor and the radio in her apartment, and then went back to running the elevator.

When I was leaving the hotel on the first day of my investigation, there was a woman at the entrance to the

•

hotel who said, "You have to do something about this. This is terrible. Seven women have died in this hotel."

I asked the detective who had responded from 4th Homicide whether or not this was true—that they had seven people killed over here? He said, "No. There may be seven homicides on the block, but it never happened in here."

My thought at that time was if they had seven women killed in this hotel in a year's time then we were dealing with a serial killer, because the only other place that would have that many people dying in a short period of time would be a hospital.

He assured me that that was not true. He either lied, or he misquoted the information that he had at hand.

Calvin previously had been in jail for robbery and drugs. But when he got out he took to murder, and it was middle-aged white women that were his victims. I think he had a hatred of white people that drove him to pick these people as victims, plus the fact that they could not defend themselves and that if he so chose he could engage in a rape or a sexual act with them. He would also take money or goods from them, but he couldn't take much from any one of them, because most of them were low-income people or on welfare in this particular hotel.

Miller was the only one that he had bothered to set on fire. So, my portion of the case was strictly centered on her.

Many of the murdered women had initially been given accidental causes of death by the police department and by the medical examiner's office. The reason that occurred was that after he killed these women they would not be discovered until either the odor that they created by the rotting flesh was too much for someone to take any longer and a complaint was made, or when their rooms were entered because they had not paid their rent. In their state of decomposition they did not show any visible signs of trauma. Because of their history with alcohol and possibly

•

drugs, the signs that they did have initially mimicked death due to substance abuse.

I believe, however, that several of the deaths, up until Miller's homicide, were declared to be suspicious.

There was a porter that worked in the hotel who was arrested for one of the cases, but after Calvin confessed it turned out that he was an innocent person.

Calvin had killed a woman and thrown her body into the air shaft. The air shaft terminated on the second floor. She went down at least four floors and landed headfirst. Her head protruded through the ceiling into a second-floor room.

The occupant of that room came downstairs to the desk clerk and made a complaint. The nature of his complaint was that he wanted someone in charge to "make this stop," and when he was questioned as to what it was that he wanted stopped, he said, "A head was sticking through the ceiling, and blood was dripping down into his sink." He didn't mind it for the first couple of days but now it was going into the fourth day. He didn't think that it was unusual that somebody's head was sticking through the ceiling, but the fact that she was bleeding all over his sink got to be a little bit annoying.

Well, the cops came around, and they declared that particular death a homicide. I think it was rather obvious to them, because there were enough marks on the body to indicate that there was some foul play, and not too many people jump into air shafts to commit suicide.

After Calvin confessed, all of the cases were reopened. Some of the bodies were exhumed, and each one of them was declared to be a homicide, because there was greater effort put into it by the medical examiner's office. They checked the small bones in the necks of the victims, and they found a number of them to be broken or crushed indicating that strangulation was the manner in which he killed all of them.

•

Calvin was not a big man. He was a medium built individual, about 5'8 or 5'9.

He was subsequently tried for all of these crimes. I testified on my portion of the case. I told them when I arrived on the scene I had to start my investigation somewhere, and when I asked him a question he recoiled a little bit. While I was making this statement up on the stand, they asked me, "Do you see that individual that you spoke to that day in the courtroom right now?" All during the course of the trial Calvin kept his head down on the defense table with his arms over the top of his head.

I said, "Well, knowing who you have sitting at the table, I would say that is him, but I can't tell you for certain that is him, because I can't see his face. I don't know if, in fact, the person that you have sitting there is the person that I spoke to. And, unless he picks his head up I can't tell you if I see that person in court."

The defense attorney said, "We still stipulate that the person sitting at the table here, Calvin Jackson, is in fact the same person that the marshal spoke to."

That exchange was strictly to establish his presence at the scene. They were not challenging the fact that he was responsible for the homicides. What they were trying to do was establish that he was not mentally competent when he did it—that he was crazy. They were doing an insanity defense.

Calvin was convicted of all nine homicides. Considering the number of people he killed, the case didn't make the headlines, but it did make the paper.


I also worked the Puerto Rican Social Club fire up in the Bronx. That was a relatively simple fire. A container of gasoline, less than a gallon, was spread about on the stairway leading to the second-floor level of the club. The remainder of the gasoline was just thrown on to the

•

ground floor of the entranceway, and a match used to ignite it.

The individual involved in setting that fire, José Antonio Cordero, utilized the services of several young men. They set three fires that night. The first was the social club. Then they burned a van that was owned by the boyfriend of the girl that Cordero was jealous of. And they set fire to a car that belonged to people who had stored some of Cordero's belongings for over a year. They had been pressed by the sanitation department to get them out of their basement; and had asked José to come and take his stuff out, but he chose not to do so. So after a while they had put his stuff out on the street, and it had been taken away by sanitation.

Cordero could have just as easy set fire to their house rather than to their car. By that token he probably saved the lives of the individuals living in the house.

The investigation took a long time because we didn't know where the problem was. It was strange, everyone I spoke to did not tell the truth. We would go off and investigate the statements that they made to determine their validity, and we would find out that they were not true. When we went back and confronted them, they all said, yes, they did lie. When we asked them why, they would say, "I don't know, I don't know."

It was amazing to see how many of the people in the social club or members of their families were involved in some form of violence prior to this tragedy occurring. So, every one of them was suspect. We had to investigate to see whether or not it was in retaliation against some of the individuals that were in the club, for previous homicides.

There were also Cuban mafia ties involved in there, and we were looking to see whether or not there was a gambling aspect to it.

There was much work that was done that, in the end, turned out to be unnecessary, but in any investigation you don't know what is necessary and what is not until the completion of the investigation.

•

In the beginning we had 10 fire marshals and 70 detectives assigned to the investigation.

I developed two people that were supposedly on the scene, but the story that they told just did not jibe with other information that we had. One of them had gone to Florida, but we got him back.

The other individual Donald Washington and I interviewed on a Saturday. The initial interview was really dynamite. It looked as if we had detected the individuals that were responsible for making this fire. I could have gotten him to confess to setting the fire in ancient Rome.

Then he told us how he went to Times Square, and Forty-second Street. He used to wander around down there. Below the Forty-second Street area it starts to become desolate. It's a very commercial area, and you quickly run out of people. Well, he went down about two blocks, when he saw an individual sparkle in, on top of a sewer cover, as they do on "Star Trek" when the crew is beamed in the transporter.

This individual was beaming up from an underground location, and he had a box that was attached to a strap around his neck. This was the box that he used to beam himself in and out of reality. He engaged this individual in a conversation, and he acceded to the individual's request to accompany him to his underground lair. All he had to do was hold the guy's hand, and they guy sparkled them both into this underground location where there were tanks and planes and other things.

I don't know if there were girls down there, but there were many weird things that he saw. He spent some time down there, and then the guy sparkled them back on top of the sewer cover on Seventh Avenue and Fortieth Street.

This entire conversation was being tape-recorded. I guess there was a little bit of viciousness on my part, because I had spent a number of hours getting the individual up to the point where he told me about beaming in and out. I just packaged up the tape, after I concluded my

•

interview with him. I put the tape into an envelope and attached a cover letter, and sent it, via messenger, to headquarters.

Mike O'Connor was the chief fire marshal at the time. He received the tape on Monday morning. In the cover letter I told him it was imperative that he listen to tape in its entirety, because I think we may be on to something here.

I was told later on, that he sat there with Tommy Flannagan and listened to the tape. The witness was coming across well, until he got to the sparkling in and out. O'Connor just said, "That son of a bitch."

Needless to say, we stayed in the Bronx doing a lot of checking. We worked around-the-clock on it initially, but it got cut down to two tours from 8:00 in the morning until midnight. Then we got down to just working during the day. And instead of seven days a week, we were down to five days a week.

The individual who had been locked up for grand larceny-auto, and who had given the district attorney's office a tip about this case after getting arrested was still in jail. I assume he was brought back up for sentencing on the GLA, and he complained again, "Hey, I gave you the fire, what are you doing? What do you mean you're going to send me to jail?"

The DA's office gave the tip over to the police again, and said, "Check out this guy. He is saying these people are responsible."

An individual was then picked up, and he gave up some hard information. As a result of that, I picked up a second individual who was a teenager. When I spoke to him he said yes, he knew about this. He was walking around with that information in his pocket all of that time and couldn't understand why we weren't taking any action. He was originally given an offer to join Cordero's group and participate in the burning, but he chose not to. However, he did name all the participants.

•

As a result of that, some further investigation was done, and it confirmed the fact that these people were in fact the ones involved.

They grabbed a 17 year old who was just on his way into the navy. They sat him down and he confessed to his participation in it, and further implicated the other two people.

Then they arrested Cordero. However, the other kid had gone to Puerto Rico. His nickname was Beansy.

Both Jack Lovett and I went over to the gang squad that was in the Four-O precinct. The sergeant in charge of that unit is now a professor out at St. John's University. He supplied us with the actual name of this Beans kid. It was Francisco Mendez.

We went to Francisco's house with shotguns, because there was information that Cordero's sons were out on the street looking for the other participants, and they were going to kill them.

We knocked on the door and announced our presence and our purpose. We told him, "Open up, we have the place surrounded." The classic *you're surrounded, come out with your hands up*, but nobody was answering the door. Then we heard what sounded like movement inside the apartment. So, we kicked the door open and went inside.

It turned out that the noise was made by a Doberman pinscher. We had to secure the dog in the bathroom. But there was nobody there.

Everybody except Jack Lovett and me took up. They were waiting for us in the street while we were making arrangements with the superintendent to secure the apartment. All of a sudden, Francisco Mendez's sister comes running into the apartment, and we grabbed her.

She subsequently took us to her relatives in New Jersey. They all knew about the fire and this kid's involvement with it and told us where he was in Puerto Rico.

Tommy Flannagan and John Regan flew down to Puerto

•

Rico, along with a member of the Seventh Homicide, and in a couple of days got Francisco out of the hills.

We all went out to La Guardia Airport to meet their plane. Then we brought Francisco over to the Four-Eight precinct in the Bronx, where someone from the district attorney's office interviewed him. He confessed his entire involvement in it.

There were three of us together one night. We were working in what was the first Bronx arson task force. I had to go up and interview a witness who lived on Beaumont Avenue, in the Italian section.

When I was nearing Beaumont there were two cars directly ahead of me. They made a screeching left-hand turn on to Beaumont, and pulled up sharply. They double-parked in the street, which was not that unusual for that neighborhood. I guess it's not unusual for New York City.

I could not find a parking spot so I proceeded across 187th Street. I found a spot. As I'm pulling in, I observed that the people were now out of those double-parked cars. The doors were open, and they are yelling in a foreign language. I didn't know what language it was at the time.

I see guns in the hands of two of the males that were there. The one female is still in the car. The two males come off the sidewalk. One of the males takes a black automatic and places it into his waistband. He gets into the lead car. The second male takes another automatic and places it into his right-hand jacket pocket. He gets into the second car. Neither of these people appeared to be law enforcement officers.

I decided that I am going to apprehend them, and investigate who they are, and what they are doing with guns. As I get out of the car I had informed the two other men who are with me that the people had guns and we have to take them.

•

We started heading over towards the car. I did not really relish the thought of going to the first car and asking him to stop, and having the guy in the second car being able to take a shot at me. That became moot, because as we got near them the first car pulled away in a very rapid fashion. The second car, however, remained at the scene.

I approached the driver, pointed my revolver at his head, and told him I was a police officer, and not to move his hands, not to do anything. He's under arrest. I wanted him out of the car. I wanted many different things.

This individual takes his right hand and puts it down to his side where his jacket pocket would be. I then see the automatic pistol come up in his right hand.

I was standing behind the driver's side of the front seat. When I saw the gun coming up I stepped back another couple of steps. I heard an audible *click*. Then he got very flustered and he put the gun back down. I couldn't tell where the hell the gun was going, but at that point I threatened to kill everybody in the car.

I was telling myself to shoot him, but I guess because of my upbringing, I did not shoot because I didn't see the gun anymore. But I was in a position at that time where if he came out with the gun, and gotten it back into working order, he would get a shot off and probably hit me, but I would hit him. One of my partners was on the other side of the car, and the other one was at the rear of the car. So, we had them covered. We ordered everybody out of the car.

It was a very tense period of time until he got out of the car, and turned around and faced the automobile, because that's what I wanted him to do.

We started taking everybody else out. I still did not see the gun. We made them all back out of the car. So, even if they come out with the gun they were going to have to turn around before they could shoot, and hopefully we would get a shot off before they did.

We had all of these individuals lined up around the car.

•

It turned out that there were four of them. We had two over the trunk, and one on each side.

I then sent one of our guys back to our car to call for some police assistance.

The cops go to Belmont Avenue. We were standing there for what seemed like hours, and that gun I'm holding in my hand is becoming very, very heavy. It starts falling down at the muzzle, and I have to keep propping it back up. That's not the first time I've experienced that. The time gets exaggerated.

I send my guy to call the dispatcher again, and tell him that we are still waiting for the police.

I am keeping my eyes on the prisoners we have in front of us, when in my peripheral vision something moving comes to my attention. I turn around and look. Heading towards us is the individual who got into the first car, and who put the gun into his waistband. I now knew that I had an armed individual coming towards me. I tell my two partners to watch everybody.

I approached this guy who has his hands in full view. He is trying to find out what we were doing with his friends, and he is mad that we're holding them up. He didn't know who we were. That turned out to be just a sham, because I put him up against the car, searched for the gun, and came up empty. But I knew he had it. So, I know that that gun is down in the car that's down on the corner. But there is no way to go down there and get it. I figure, assuming that he was the only one in that car, it will still be there after we get assistance.

I put him up against the car alongside of the individual who pulled the trigger on me, but he kept coming off the car and saying things to me like, "Why are you doing this? What did we do to you? Why are you picking on my friends?"

I told him, "Stay on the car. Don't come off or you're going to get shot."

He says, "You can't shoot me. You can't shoot me."

•

I really couldn't shoot him, but I had the hammer cocked on the revolver. I was afraid if I clocked him with the revolver that it might go off. So, I gave him a left hook to the back of the head to get his attention and that did it. It kept him on the car, but it really pissed him off. He threatened to do things to me, not only then, but also later on. And, while I was doing that, my peripheral vision again picked up some movement.

This time it was a *second* individual who was coming down in a very stealthy fashion and he was staying close to the outside of the cars that were parked alongside the street. I saw that in his right hand, down by his leg, he had a revolver. I took up a combat position, and I hollered at him, "Police, drop the gun!"

The marshal on the other side of the car began walking towards this individual with the gun in his hand, even though I was yelling, "He's got a gun! He's got a gun!"

The marshal heading towards him was making no effort to take cover, he was totally exposed. I feel that he never heard me, but he was just so preoccupied. In these tense situations you get tunnel vision and you also get tunnel hearing.

I knew that I could shoot this guy, and I would hit him, but I didn't know if I would neutralize him. I thought that he would shoot at the marshal that was totally exposed and coming towards him, because he was the closest person to him.

The individual with the gun saw me, and saw this other person coming towards him. He became confused and really didn't know what to do. Then he dropped the gun, and attempted to push it underneath the car that he was standing alongside of. I did not have to shoot. He gave up.

We arrested all of these people and took them into custody. We ended up with three guns and six people. The gun I was looking for was in the female's pocketbook in

•

the backseat. They were Albanians. They threatened to get me.

The reason that they were there was that the one guy who pulled the trigger on me had an argument with his brother, who had said something nasty to his wife. Because of honor, he was going to shoot his brother.

The individual who had come back, and was acting as the decoy to distract us, while the other guy was going to come down and shoot us in the back, was the worst of the bunch. He made veiled threats, that he was going to meet me on the street someday and he would take care of me.

Knowing the Albanians as I do, I decided that he was probably going to make good on his threats somewhere along the line. When I had to go to court with him I would bring along another marshal who would not come into the courtroom with me. I'd have him sit in the back of the room. His job was to see that I didn't end up in a casket.

The worst Albanian's lawyer was a state senator who ended up getting locked up, indicted, and sent to jail himself.

The worst Albanian just skipped, and as far as I know there's still an outstanding warrant for him.

All the other five people were indicted, and all of them were sentenced for one thing or another.

We recently had the Happy Land social club fire. This man was going out with the girl who worked as a coat check girl there. He had an argument with her because they split up, and he wanted her back.

She told him she didn't want to see him anymore, and he said well, he didn't want her working there. She said she was going to continue working there. He said, well, you won't work here anymore.

He was in the club having a drink, and he became ob-

•

noxious. Evidently he didn't have the money for another one, and the bouncer threw him out.

He had to borrow a dollar to go and buy the gasoline. He got it in a plastic container, and came back with it. When he came back we know that he was observed, and she was alerted to the fact that he was back. She saw him come into the foyer, and she left the coatroom.

He took the gasoline and he spread it around the entrance foyer. Then he stepped outside. He put the gas container down next to the building. Then he lit a match and threw it in the foyer, and then he lit another match and threw it in. The fire took off and so did he.

He went home, and he told people at home that he had set the fire. He thought people got hurt but he really didn't know how many people got hurt. However, he did, later on in the day, because it was all over the radio.

The fire itself really didn't amount to too much as fires go. It was a very easy investigation relative to the physical examination on the point of origin and the causation of it.

All of the other aspects that went along with the fact that 87 people died as a result of this fire, and mobilized just about every agency in the city to come up there, lent itself to a major screwup, but it did not work that way. Everything worked very, very smoothly. There was total cooperation between the police and the fire marshals.

Anything that I requested, I got. I needed a helicopter to fly an accelerant-detecting dog to the scene. The New York State division of fire prevention and control has two of these dogs that we can access whenever we have a need to. We had never done so before, but I felt that I wanted the dog down there and I asked for a police helicopter to go up to Albany to pick up the dog. I did not know that fire prevention and control had an agreement with the state police that whenever they had a request for the dog the state police would supply a helicopter to fly the dog to whatever location it had to go to. So, I had a police helicopter down on the street with the engines running

•

and a state police helicopter landing right alongside of it with the dog from Albany.

This individual was apprehended, and he made an admission. Everything that he said corroborated the findings we had in our physical examination.

The case is presently in litigation. Where it's going to go, I don't know. There has been some talk that maybe they will use an insanity defense.

As the chief fire marshal I see the future as very, very bleak. The fiscal constraints that have been placed upon the entire city in general, and this bureau especially, bring us down to a point where I am going to have a hard time operating.

I know that I can't implement some programs that I would like to, and that I feel would help us make the city more fire-safe. We are going to have a hard time just maintaining the status quo and keeping up with what we have.

Arson is totally underreported. And there are a number of reasons for that. The operating force—via ignorance because they don't have the training to detect the arson—is one of the reasons. The others are apathy and empathy. There are people out in the field who say, who really gives a shit? Nothing is going to be done about this. So, why bother reporting it? There are other people out there, maybe even the same individuals, who say the fire marshals are working too hard already. So, why give them this case? Let us just forget about it.

Unfortunately it's a catch-22 situation. If you don't report it, it never happened. If it never happened, you don't need anyone to investigate it, and because of the way they measure productivity, if we don't make the run then we don't get a mark, and there is no need for conducting the investigations. Meanwhile arson is still going to go on.

•

With the bleak economic picture, I know that arson is going to increase.

Will it ever go back to the way it was in the sixties and seventies? It will get back to that rampant burning if there are not enough watchdogs out there.

In other jurisdictions, fire investigation is a stepchild. Fire departments want it, but they do not want to spend the time, energy, and money to do it the right way.

Police departments want it, because they will be able to increase their budget for allocating personnel and resources to doing this. Once they get the personnel and resources to do it, then they utilize the personnel and resources in other areas, and if the arson begins to increase too much, then they just stop reporting it. They will ask themselves, "Can we solve this case? No, we can't solve it. Then it is not an arson, it is a fire of undetermined origin. When we can *solve* it, it's an arson." They will have a very high clearance rate, but you end up with a jurisdiction that has a very low arson problem. Maybe it will have a very large fire problem, but it will certainly have a very low arson problem.

You will also find that with homicides. Many people die, but often the cause is undetermined. If it can be solved, it will be a homicide. If it can't be solved, then it is just an undetermined cause of death.

•

\ spent 22 years in the FDNY fighting thousands of fires, and at the time I never thought that anything could be more exciting or rewarding. However, being a fire marshal is the most creative and challenging work I have done. After seven years as a marshal, I no longer directly deal with fires out of control; instead I deal with people who are out of control. Many of these people are educated and affluent.

In the last week an estranged husband attempted to burn down the home of his pregnant ex-wife by pouring gasoline under the gas tank of the family car parked in the driveway. We arrested him when he returned to the scene. The reason for his behavior was anger.

A few days later we responded to the fire bombing of an occupied restaurant on a busy street. The motivation was anger.

Then we had a suspicious car fire in an afflu-   ent neighborhood. There was a strong smell of gasoline in the passenger compartment, along with a badly burned container. We got two fingerprints off the container that belonged to the ex-girlfriend. The reason for the fire was anger.

We had a doctor's kid who was set upon by some other kids. He retaliated by throwing a Molotov cocktail against the apartment building of the kids who beat him up. I asked him, what was he thinking at the time he threw the fire bomb? He replied, "I was angry."

The frightening part is that all of these acts of arson were premeditated. It's a bad trend. If we don't put a stop to it, more civilians and fire fighters will be killed and injured.

Fire Marshal Charles Wagner

We had a number of arson fires on 140th Street, which is in the 14th Battalion in the South Bronx. This one day, Tom Flanagan and I were patrolling in that battalion when we heard fire marshals Thomas Russo and Rafael Greniela calling on the radio for assistance.

Russo and Greniela had stopped to question two suspicious guys. One of the guys started running. Greniela went after him. Russo put this other guy, who was known in the neighborhood as Ralphie, up against a door to frisk him. Ralphie suddenly turned around with a gun in his hand. He took Russo's pistol from its holster. Then he took off.

We worked with the detectives and the anti-crime cops from the Four-O Precinct to put a lot of heat on the block. The next day we got a tip that Russo's gun was in a Dempsey dumpster at Cypress and 140th Street.

The word was out that Ralphie had gone to Puerto Rico. On July 12, 1975 we went to his girlfriend's apartment to get a photo of him. Greniela and Russo asked her if they could come in. She said, "Sure." I waited in the hall.

All of a sudden I heard rapid gunfire.

The girlfriend and a couple of kids came screaming out of the apartment.

I went running in. I saw Russo lying on the floor. His

•

jaw was down to his chest. He couldn't speak. I didn't see Greniela.

I asked Tom, "Where was the perp?"

He looked toward a wooden partition.

I yelled at Ralphie to give himself up.

He said, "If you want to see your partners alive, get out of here, now!"

I went into the hall and got a marshal by the name of Phil Murtha on the Handie-Talkie. I told Murtha to call for an ambulance and a police backup.

Ralphie appeared at the apartment door in his boxer shorts. He was pointing a gun at me. He said, "I told you to get out of here if you want your partners alive."

I backed down the stairs.

Then I heard gunfire from the roof. Fire marshals Tom Flanagan and Tony Kiesel were trading shots with him. They later told me that Ralphie jumped out of the bulkhead door wearing only his boxer shorts and a gun in each hand.

I went back into the apartment. I found Greniela in the bedroom. He was lying in a pool of blood. He couldn't breathe. I started giving him mouth-to-mouth resuscitation.

Rescue 3 was returning from a run when they heard the call for help on the radio. They came in with a Stokes basket and took Greniela to Fordham Hospital.

The cops, who came in on the 10-13, took Russo to Fordham in a radio car.

Russo spent six hours in the operating room. His heart arrested twice on the table. He is alive today, but he lost 50 percent of the movement in his neck.

Greniela was shot in the spine. He was later transferred to Bellevue Hospital. He is paralyzed for life from his chest down.

After his wounds healed, Ralphie was convicted of attempted murder and armed robbery. There is a good chance that he soon may be out on the street.

Deputy Chief Fire Marshal Mike DiMarco

•

*  *  *

I spent two years as a correction officer. Then on June 6, 1962 I entered the fire department. After training I was assigned to "the garden spot of the Bronx," Engine Company 73. It was also known as "La Casa Ca-Ca." We were in quarters with Squad 2 and Ladder 42, "The Elephant House." The dichotomy could not have been any sharper: The truck was spotlessly clean, while the engine was "lived-in." I spent the next 16 years there, working with the best collection of nuts and fools, as the Bronx burned down around us.

On June 28, 1978 I became a fire marshal. I was fortunate to be assigned to Brooklyn, the busiest borough for the Bureau of Fire Investigation.

With 16 years of very active fire duty, I thought I could read a fire and find the source without too much difficulty. I found it to be the complete opposite. Luckily, my training officer challenged me at every job. He made me think and explain my conclusions. Then he proceeded to show me the truth.

The fire department has been good to me. I have had the best of both worlds. I was a fire fighter at the peak of the fire storm, and a fire marshal involved in reducing the rate of fires.

Now, with 28 years on the job and retirement more of a reality, I am not looking forward to the day when I will leave behind the long and deep friendships that I live with every working day. They will be sorely missed.

I will not miss digging through the slop and the slime and the ashes, but I will miss the joy of a successful prosecution of someone who tried to hurt my brother fire fighters.

It was very hard to leave Engine Company 73. They were as close to me as my family is. I know it will be even harder to leave the Bureau of Fire Investigation.

Fire Marshal Arthur Crawford

•

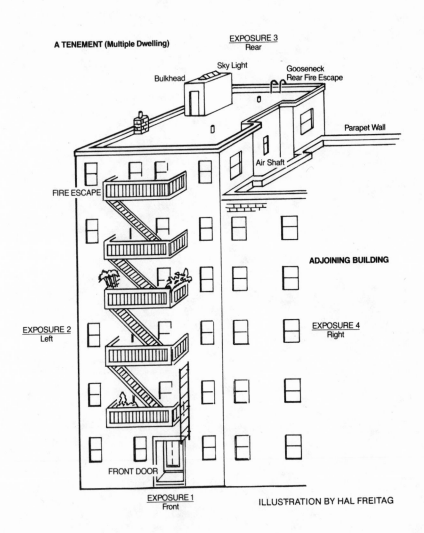

A TENEMENT (Multiple Dwelling)

EXPOSURE 3
Rear

Bulkhead

Sky Light

Gooseneck
Rear Fire Escape

Parapet Wall

Air Shaft

FIRE ESCAPE

ADJOINING BUILDING

EXPOSURE 2
Left

EXPOSURE 4
Right

FRONT DOOR

EXPOSURE 1
Front

ILLUSTRATION BY HAL FREITAG

# GLOSSARY

ACCELERANT     Something, usually a flammable liquid, that is used to increase the spread of fire.

ADA     Assistant district attorney.

ALL-HANDS     A fire in which the entire first-alarm assignment is working.

ALLIGATORING     The charring of wood as a fire burns. Large, shiny squares indicate a quick, intense fire.

ALPHABET LAND     Avenues A, B, C, and D on Manhattan's Lower East Side.

APPARATUS     Another term for fire trucks.

APPARATUS FLOOR     The ground floor of a firehouse where the trucks are kept.

ATF     Bureau of Alcohol, Tobacco and Firearms. It is a division of the U.S. Treasury Department.

•

| | |
|---|---|
| BATTALION | A group of engines and ladder trucks that cover a particular geographical area, under the command of a battalion chief. |
| BLACK BEAUTIES | Pills containing a combination of Benzedrine and Biamphetamine—which are both stimulants. |
| BOMBERO | Spanish word for fireman. |
| BOX | One of 16,000 alarm devices on the city streets. Every alarm is assigned the number of the nearest box. |
| CHICKEN HAWK | A procurer of young boys for homosexual men. |
| CLASS I | The highest level of heroism award in the FDNY. |
| CODE | Used in conjunction with other radio signals to indicate the degree of severity. Code-one is the most severe. |
| COLLAR | Police jargon for an arrest. |
| COMPANY | The fire fighters assigned to each ladder truck, engine, rescue, or fireboat. Each company is commanded by a captain and 3 lieutenants and has approximately 25 firemen assigned to it. |
| COP A PLEA | To plead guilty to a lesser charge with a lighter sentence, rather than risking a longer sentence if convicted in a trial. |

•

| | |
|---|---|
| CYANOTIC | Exhibiting *cyanosis*, a bluish discoloration of the skin resulting from insufficient oxygen in the blood. |
| DOA | Dead on arrival. |
| DROP A DIME | Short for dropping a dime in a pay phone to inform the authorities about someone's criminal activity. |
| ENGINE | The apparatus that carries the hose lines and pumps water. |
| FELONY | Any offense punishable by death or imprisonment for a term in excess of one year. They are designated as A, B, C, or D, with A being the most serious. |
| FINGERING | Splash effects of flammable liquids. |
| FIRE TRIANGLE | Heat, fuel, and oxygen—the three components of fire. |
| FIRST DUE | The first assigned truck or engine company to a particular location. |
| FLASHOVER | The stage of a fire when a room or other space becomes heated to the point that flames flash over the entire area. |
| GLA | Grand Larceny—Auto. |

•

| | |
|---|---|
| GRAND JURY | A group of individuals impaneled to determine if enough evidence exists to indict someone for a particular offense. The standard is that there is enough credible evidence to prove it more probable than not that the individual committed the offense. |
| HALLIGAN TOOL | An all-purpose steel prying tool invented by Huey Halligan, a former member of the FDNY. |
| HUNTLEY HEARING | A pretrial hearing to determine the admissibility of certain statements made by the defendant that may be used as evidence against him or her. |
| HOOK | A 6-, 10-, or 20-foot pole with a pointed metal tip and a short hook several inches below. |
| INCENDIARY FIRE | An intentionally set fire. |
| IRONS | The forcible entry tools, particularly the axe and the Halligan tool. |
| LINE | Short for hose line. |
| MARICON | Colloquial Spanish for sissy, pansy, homosexual. |
| MARK | A murder contract. |
| MASK | A self-contained breathing apparatus that consists of a face piece, regulator, air tank, and backpack. |

•

| | |
|---|---|
| M-80 | A very powerful firecracker. |
| MIRANDA WARNING | Prior to any custodial interrogation a person must be warned that: 1. They have a right to remain silent. 2. Anything they say can and will be used as evidence against them in a court of law. 3. They have a right to have an attorney present during questioning. 4. If they cannot afford an attorney one will be appointed for them by the court at no cost to them. |
| MOLOTOV COCKTAIL | A bottle containing gasoline with a cloth protruding from the neck that serves as a fuse. |
| NAGRA | A very high-quality recording device used by law enforcement agencies. It will play up to three hours, but the person wearing it usually doesn't know how to turn it on or off. The tape requires special equipment to be played back. |
| NEW-LAW TENEMENT | A brick apartment building built after 1901. Usually about six stories in height (see diagram). |
| OLD-LAW TENEMENT | A brick apartment building built before 1901. |
| PDU | Precinct Detective Unit. |
| POLE | The brass pole on which firemen slide down to quickly get to the apparatus floor. |

•

| | |
|---|---|
| PREDICATE FELON | A designation given to an individual convicted of a prior felony that increases the sentence on the present charge. |
| PRIMARY SEARCH | The initial search for victims conducted by fire fighters while the fire is going on. |
| PROBABLE CAUSE | A set of facts and circumstances that would induce a reasonably intelligent and prudent person to believe that a particular person had committed a specific crime. |
| PROBIE | Probationary fire fighter—e.g., a beginning fire fighter. |
| RACKING | The cocking of a pump shotgun. |
| RICO | Racketeering-influenced corrupt organization. |
| RIKER'S ISLAND | New York City's primary municipal prison. |
| SCOTT AIRPACK | An early type of self-contained breathing apparatus (mask) manufactured by Scott. |
| SECONDARY SEARCH | A thorough search for victims by fire fighters, usually when the fire is knocked down or under control. |
| SOP | Standard operating procedure. |
| SQUAD | The radio designation of a fire marshal car, e.g., Squad 41 Alpha; also another way of referring to a PDU. |

•

| | |
|---|---|
| SQUAD 4 | A floating company of fire fighters that provides extra manpower. They carry additional tools on an engine and have a very wide and active response area. |
| TAKE UP | FDNY jargon meaning to go. The implication is that after a fire you gather up your hose lines, ladders, and tools before leaving. |
| TAXPAYER | A row of stores one story high, though occasionally there will be a second story of offices. |
| 10-4 | FDNY radio signal indicating that the message has been received, understood and will be complied with. |
| 10-41 | The radio signal indicating a suspicious fire and fire marshal investigation is required. |
| 10-45 | The radio signal indicating a severely injured (Code 2) or dead (Code 1) individual found at a fire. Sometimes used as jargon for a dead person. |
| 10-75 | The radio signal given by the first arriving unit indicating that a working fire or emergency exists. This signal tells the dispatcher to fill out the first alarm assignment and start a rescue company on its way. |
| TRUE BILL | An indictment handed down by a grand jury. |
| WISEGUY | A member of organized crime. |

•